Natural Knowledge in Preclassical Antiquity

"Then Zeus decided to restrain his own power no longer . . .
the bolts of lightning and thunder flew fast and thick from
his powerful arm"
(Hesiod, *Theogony*, book 10).

▼ ▼ ▼

Natural Knowledge in Preclassical Antiquity

Mott T. Greene

The Johns Hopkins University Press
Baltimore and London

BL435
.G74
1992

The Johns Hopkins University Press
701 West 40th Street
Baltimore, Maryland 21211-2190
The Johns Hopkins Press Ltd., London

∞

The paper used in this book meets the minimum requirements of American National Standard for Information Sciences—Permanence of Paper for Printed Library Materials,
ANSI Z39.48-1984.

Frontispiece: Volcanic lightning at Surtsey, in the North Atlantic, December 1, 1963. From *Surtsey*, by Sigurdur Thora-
rinsson. Copyright © 1964, 1966, by Almenna Bokafelagid.
Used by permission of Viking Penguin, a division of Penguin
Books USA Inc.

Library of Congress Cataloging-in-Publication Data

Green, Mott T., 1945–
    Natural knowledge in preclassical antiquity / Mott T.
Greene.
        p.   cm.
    Includes bibliographical references and index.
    ISBN 0-8018-4292-1 (alk. paper)
    1. Knowledge, Theory—History.   2. Natural history—
History.   3. Science, Ancient.   4. Philosophy, Ancient.
I. Title.
B185.G74   1992
121—dc20   91-26952

To the Memory of Roderick MacArthur

▼ ▼ ▼

Contents

Acknowledgments

My thanks to the John D. and Catherine T. MacArthur Foundation, and to Ken Hope, the director of the Fellows Program, for the MacArthur Prize Fellowship awarded me in 1983. The fellowship and the spirit in which it was awarded provided all the means and some of the inspiration to write the book. The dedication of the volume to the memory of Roderick MacArthur expresses this gratitude more directly.

I thank Jo Leffingwell for her suggestion of a possible connection between Soma and ergot, and for emphasizing the importance of the *Satapatha Brahmana* in this regard. She also provided information on multiple channels in meandering streams, and discussed with me many of the issues raised in other essays.

W. D. O'Flaherty commented on an earlier draft of the essay on Soma, and G. E. R. Lloyd on the essay on Thales. Thomas G. Palaima was helpful in rethinking the context of the work on Egyptian fractions. These were all useful comments, but do not constitute endorsements. The Ancient World Symposium members, my colleagues at the University of Puget Sound, also gave helpful advice—especially David Lupher and Bill Barry. Peter S. Greene, M.D., suggested the possibility that lamb's wool could function as a molecular sieve for toxic fractions of ergot, and answered other biochemical inquiries as they rose.

My thanks to Hannah Wiley for a perceptive reading of an earlier draft, to Sheryl Schmit for help with the bibliography and index, to Jean Brooks for careful manuscript preparation under tight deadlines, to Ann Waters for her helpful copy editing, and to Robert Brugger of Johns Hopkins University Press

Acknowledgments

for his interest in the project. I also thank the Press's anonymous reader for his or her suggestions.

Thanks to my niece, Heather Greene, for watching Annie in the summer of 1985 while I wrote part of this.

My special thanks to the anonymous donor of the John B. Magee Distinguished Professorship in Science and Values at the University of Puget Sound, whose generosity provided me with a permanent academic home and research support and writing time to bring this work to completion.

Introduction

The introduction is the author's last stand over the body of the text and his last chance to say the things that (somehow) still remain unsaid with a work already completed. From this vantage, a long introduction is an admission of defeat, or a sign of bad conscience, or both. I shall therefore be brief.

These essays are the result of some time away from my usual preoccupations—studying the history of modern earth science and making my living as a college teacher—afforded me by a MacArthur Fellowship awarded in 1983. I decided to spend the first part of that five-year fellowship "cleaning out the attic"—that is, pulling out of the recesses of my brain some ideas that I wished someday to pursue, but had hardly expected to have the leisure to follow up so soon. I should have seen from the outset, but did not, that the project would easily and rapidly grow far beyond a little mental housecleaning and end up being the labor of several years, which it is. Indeed, several of these essays might well have appeared, with more ample documentation, as individual books, and might have done so if I had not other research plans already under way, and indeed interrupted, by the length of these studies.

This book is a series of meditations on the relationships between mythology and natural knowledge both in antiquity and in the present. I have used the phrase "natural knowledge" rather than "science" in my title because what we generally think of as "science" is only one of many ways of getting, storing, and conveying knowledge about our natural surroundings. To call what the Greeks, the Egyptians, or the Vedic Aryans did in their encounters with the natural world their "science" prejudges and limits those encounters, and carries

Introduction

misleading connotations (of progressive research traditions, of controlled observation and experiment, of mathematical ordering of nature) quite peculiar to our own, modern forms of natural knowledge. I have tried to show that there are a number of puzzles in prehistory, ancient history, and the study of mythology that cannot be unlocked by the literary use of literary evidence, even when supplemented by archaeology. Some detailed and accurate knowledge of modern natural science is required as well to make sense of the mythic material, insofar as this mythic material also is detailed and accurate material about the natural world, albeit given in an unfamiliar form.

The "ancients," as we generically refer to everyone from the time of Plato back to *Homo erectus*, lived and worked outdoors, and did most of their thinking there as well. Because we have come indoors in the last hundred years, much that was obvious to earlier generations of scholars about ancient mythology is no longer obvious to us. That myths are to a large extent stories about nature has passed in the last few generations from something that "goes without saying" to something that "cannot be said." When Darwin wrote *The Origin of Species* in 1859, he introduced the volume with a long disquisition on the breeding of domestic animals, as a way of leading his audience to the difficult terrain of natural selection through a familiar backyard (and barnyard) path. Today this material is as foreign to most readers as the subject it was meant to illustrate—and this is because we have, in the intervening century, come indoors.

On the other hand, what I offer here is not (may the gods be my witness) *interdisciplinary*. This designation, which a generation ago pointed hopefully toward integration and synthesis of disciplinary studies, now reeks of superficiality, social science, and pallid syncretism. Within the academy, *interdisciplinary* is now synonymous with programmatic, insubstantial, overtly speculative endeavors which discard not only disciplinary boundaries but canons of scholarship and rules of evidence as well. Interdisciplinary rhetoric, for all its talk about tearing down barriers, is based on a word play—substituting

Introduction

for "discipline" as specialized knowledge requiring special training, "discipline" as enforced order, unthinking obedience to superiors, and plain bondage. The *interdisciplinary* scholar playing this game sees her- or himself as transcending the intellectual bondage and the limitations of perspective of a single discipline, and sees her or his work as moving freely in the space between disciplines, rather than as an onerous submission to a double sovereignty: observing two sets of rules and two canons of evidence as governing one's research, rather than a single rule.

Strictly speaking, there is no space between disciplines. It is not that all possible areas are being studied. Rather, one cannot say anything substantive about areas into which no discipline has ventured simply because there is no research there on which to base statements. Statements about such areas must, from the standpoint of the disciplinary worker, appear as shirking, dilettantism, and fakery—and scant wonder they should!

What appears here is, on the contrary, the collation and combination of professional, specialist results from several currently distinct disciplines, of necessity, for addressing the specific problems I tried to solve. The limitations of such combinations are apparent to me. Academic disciplines are very like biological species, and develop their special characteristics, appearances, and habits over time and under strong selective pressures. Like "good and true" species, as Darwin would say, they can occasionally produce, when intercrossed with other species, hybrids of great utility, but limited fertility. A mule can work longer than a horse, and carry more than an ass (and often appears smarter than either), but it cannot produce little mules. These essays are mules in that way. They are useful hybrids intended for hard work, not little formative disciplines setting seed for progeny. That they exist at all is testament to the realities of the world of problems and solutions—not every interesting set of problems has its own discipline ready made, and not every set of problems perhaps needs or deserves its own discipline.

Introduction

I have written about a number of different subjects in this book, and from a variety of standpoints. Yet all these essays focus on the single problem suggested by my title, "natural knowledge in preclassical antiquity"—what it was, how it was held, and by whom. Therefore, I begin this book with a discussion of the general lesson to be drawn from the essays taken together.

The first essay, on the ideas of primitive man and prehistory, is a meditation on various attempts to construct a criterion that could make the appearance of *Homo sapiens* an event of radical evolutionary significance whereby our biological descent is qualitatively transformed and succeeded by a historical "ascent to civilization." I conclude therein that "prehistory" is a permanently established frame, which shifts forward and backward in time, a place where modern theorists construct origin stories about humans. In the most recent versions, these stories associate increased cranial capacity (and therefore brain size) with emergent mental capabilities, and with the purportedly simultaneous appearance in the archaeological record of fine tools, ornaments, and images. I show that such schemes illegitimately equate the scant variety of materials *preserved* from earlier periods of hominid existence with the kind and number of materials *possessed* by earlier hominids. I suggest that, in particular, edge-wear analyses of stone tools show that the full variety of later materials (hide, bone, vegetable matter) were worked not tens, but hundreds of thousands of years ago, and I conclude that there is no warrant for creating "human beings" in the recent hominid past via this approach, or for retrojecting stories of progress into the more distant biological lineage.

In the second essay, on Egyptian fractions, I make the same point again from a different angle and in a different historical context (also of high antiquity). I note that multiple theories of "lost Egyptian wisdom" have been created to eliminate a discomfiting disparity between the excellence and mathematical precision of Egyptian constructions, such as the great pyramids of the Old Kingdom, and the simple tools and cumbersome mathematics that have come down to us from the same

Introduction

time. Here I have argued that the curious "unit fractions," which Egyptians never abandoned in favor of other and simpler methods of fractional calculation, were founded in acts of practical measurement with unit-fractional weights. I concluded that the Egyptians "failed" to "improve" them because they saw the unit fractions not as abstract intellectual constructs, but as specialist tools in a culture of "craft numeracy" in the hands of the "craft literate" scribes. That is to say, they produced great achievements with simple tools skillfully used. Here I meant also to suggest that our notion that natural knowledge has some internal dynamic which is progressive, cumulative, and directed toward novelty, generality, and simplicity is not well supported by the historical record of a civilization of high antiquity—Egypt—often proposed as the formative stage in such a dynamic. I have deliberately challenged any attempt to construct the kind of "origin story of the sciences" that one finds in almost every Western Civilization textbook.

The third and fourth essays challenge the narrative of the simultaneous emergence of rational thought and of scientific investigation in Hellenic Greece, portrayed as a transition from a mythical and superstitious cosmology in Homeric times to a scientifically recognizable material cosmology in the work of the pre-Socratics of the seventh century B.C. Here I spent some time on the *Theogony* of Hesiod to make a number of points, but principally to argue that there is strong internal evidence in this poem suggesting that it contains accounts of volcanic eruptions, described as battles between giant beings. Furthermore, the content and ordering of these descriptions make it possible to identify which volcanoes are described. This leaves us with a picture of the emergent cosmic order under Zeus, described by Hesiod, in terms of historical descriptions of natural events. In the subsequent essay, on the Cyclopes, I explore (necessarily somewhat repetitively) how this original equation of events and beings could be generalized to provide useful information about where to travel and where to settle—and where not to. Herein I wished again to show the danger of retrojecting concepts, vocabularies, and organizing schemes in time and then constructing lineages out of them. These myths are not proto-science or proto- anything; they are a useful form of natural knowledge, differently held.

In the fifth essay, on Thales, I made an explicit attack on the next chapter of the myth of progressive emergence by contrasting the portrayal of Thales and his place in the Greek Miracle as the first disinterested observer of nature of whom we have record with the evidence that we have about him. In so doing I point to the Greek version of what Thales did (very different from the modern version). Using the story of the fording of the Halys River as a testing ground, I argued that attempts to employ Thales as a way of locating the "birth of natural philosophy" in a deliberate turn away from practical and manipulative attitudes about nature cannot be sustained, since such attempts force us into elaborate strategies of denial of significant portions of the record concerning Thales, which can be encompassed completely by considering him as, above all, a hydraulic engineer. I asserted that both popular and specialist accounts of Thales are quite deliberately slanted to create a pre-Socratic transition space for the story of the evolutionary explosion of Hellenic Greece, the Ur-cultural locus of Western civilization's ascendant course, and that these accounts do so by avoiding the extent and character of Thales' approach to natural knowledge. The polemical edge of this essay is deliberate, and directed against a double game I find played over this terrain: scholars claiming, when confronted with confuting evidence, to find the "Greek Miracle" an overstatement, while continuing to pass it off on generation after generation of undergraduate students via the 20-, 30-, 50-, and 100-year-old standard works through which it was originally propounded.

The connection between all of this material and the next essay, on *soma*, needs a more ample explanation. In each of the essays that preceded it, my principal approach was to apply knowledge derived from modern natural sciences and the history of science in order to recharacterize a time (prehistory), a civilization (Egypt), a story (*Theogony*, the oldest written

Introduction

Greek myth), and a person (Thales), all of which play successive and crucial roles in the story of the emergence of modern
humankind and modern Western civilization as they are most
often currently taught in universities in Europe and North
America. In each case I suggested that the prevailing story
was constructed by assuming a modern, that is to say, post-
Enlightenment attitude about the character of knowledge
about the natural world (that it is progressive and cumulative),
and then by reading the available evidence from the past in the
light of this hypothetical narrative of progressive emergence. I
suggested along the way that this situation was further complicated by a decision not to include study of the natural sciences
as part of the training of classicists, thus producing in them
either a disinclination or an inability to look for the schemes
of natural knowledge embedded within earlier periods, or to
recognize them for what they are when they see them. This
transforms the problem from some sort of nefarious conspiracy
to an artifact, at least in part, of schemes of graduate education.

In turning from Greece to Indo-Iranian antiquity, I wished
to make a related point within the ambit of a more distantly
related but not quite separate cultural tradition. I chose an
unsolved problem—the character of the Soma plant used to
induce ecstatic visions in Vedic religion—on which science
(botany) and literary study (classical philology) have labored
cooperatively for about 200 years, since the time of Sir William
Jones, at least. The relentless detail (as many readers have
informed me) of my essay on this topic is inescapable. I had
to show that the problem of exploring natural knowledge in
preclassical antiquity is not as simple as "joining humanities
and sciences together" and that I was not offering a ridiculous,
formulaic panacea for difficult problems in cultural history.
Rather, I extended my argument to include the notion that
which particular one of the modern natural sciences is applied
to problems in cultural history, and the state of development
of that science when applied, will also determine the success
of the enterprise. This is because, even if not all schemes of
natural knowledge are progressive and cumulative, modern

Introduction

physical science is progressive and cumulative. Therefore, attempts to meld scientific and humanistic study must be repeated at intervals to be successful. To do so I had actually to make a scientific argument about a classical problem, and then fit it to all the evidence available, including alternative modern schemes.

My conclusion was twofold. First, that the appropriate science for discovering the character of Soma was biochemistry, not plant morphology, on the hypothesis that Soma was the name for an active principle and not a species of plant. Absolutely central to this demonstration was the necessity to attend to the descriptions for the preparation of Soma as they are written (in relentless detail) in the Brahmanas and elsewhere. Just here, I forge the link to the earlier essays. Whether it be the description of the battle with the Titans in the *Theogony*, or the characteristics of Cyclopes, or where Thales may have gone and what he might have done, my hypotheses were produced by a reading of the mythical and sacrificial texts as narrative reports of observations and methods and not as poetical fictional creations. It is here that I have insisted on challenging the overwhelming diversion of interest in mythology away from its contents toward its forms and transformations. Such an exclusive concentration on forms must be seen to distort not mere details, but the overall function of these entities (myths) and their informative content. While it is possible to learn much about mythology by examining its formal structures, this is not a complete study of mythology. One might say by way of analogy that we may learn much about the principles of mechanics by taking apart a watch to see how it works; and yet we must also have firmly in mind that people make watches for the exclusive purpose of telling the time, and not at all to exhibit the principles of mechanics. So it is with myths. They were not made to exhibit the structures of human consciousness (though they do); they were made to transmit the information they contain.

In the last topical essay, on Plato's *Phaedrus*, I turned directly to Hellenic Greece and extended this approach and con-

Introduction

cern by enlisting Socrates himself as a figure whose aims cannot be understood apart from the specific contents of the natural knowledge of his time, and the specific means of having that knowledge—who had it, what it consisted in, how it was to be held. His medical model of education, his psychophysiological theory of soul growth and its relation to contemporary theories of vision are all necessary components in understanding his attitudes to rhetoric. This essay serves as an implicit recapitulation of my previous themes, since *Phaedrus* discusses Egypt, craft literacy, the nature of myth, and the proper aim and character of knowledge.

A number of classical scholars who have seen these essays have observed what they have called their polemical and iconoclastic tone. I have indeed been critical of the approach taken to these problems by other scholars before me. This does not mean I have no debt to them, and I wish to acknowledge that debt here. I have assumed that ultimately all of them hoped their work would find an audience beyond their own specialist colleagues, and that they would not have provided translations and commentaries on ancient texts unless they wanted people to know of them and think about them. This book would, in any case, not have been possible without the research aids produced by these scholars over several generations. If I disagree with their conclusions and interpretations, I do so on the basis of the evidence which they themselves have provided, and for this I make no apology.

Natural Knowledge in Preclassical Antiquity

▼ ▼ ▼

1

Prehistory

Prehistory is a curious designation. It appeared first in early nineteenth-century Europe as a way of describing researches into periods, however proximate in absolute time, for which there were no written records. Thus the term *prehistory* might be applied to the reconstruction of the lives of inhabitants of Wales or Scotland in the Roman period of British history, to the extent that these fell outside the pale of literacy.

Somewhat later in the century, in the terms of a scheme proposed by the Danish historian Jens Jacob Worsaae (1821–85), prehistory acquired periods of its own based on a comparative scheme of material culture: Stone Age, Bronze Age, and Iron Age, in which archaeological residues of weaponry and decorative arts came to the fore as identifying characters. Worsaae's scheme, which overlapped the chronology based on the acquisition of writing, was rapidly folded into the emerging narrative of human evolutionary development. Once the antiquity of man was established well beyond the horizon of Bible-based chronological estimates (ca. 7,000 years), prehistory began to unroll backwards in time, and new divisions appeared—the Old Stone Age distinguished from the New Stone Age (Paleolithic and Neolithic), Early, Middle, and Late Bronze Age, and so on. By the very late nineteenth century, the discovery of a wide range of human fossils and skeletal remains and associated tools led to systems of chronology that marched resolutely from apelike predecessors with a few blunt instruments to fully human ancestors possessing an abundant array of useful, complex, and beautiful tools and other artifacts.

Natural Knowledge in Preclassical Antiquity

However natural this progression came to seem and how fully documented it was, historical writers for the most part could not quite bring themselves to close the gap between prehistoric archaeology and "history." In the space between geological time and historical time, there remained a temporal province called "prehistory." Glyn Daniel pointed out some time ago in *The Idea of Prehistory* (a splendid book) that from a logical point of view "prehistory" is a misnomer, since if it were before history we should know nothing about it.[1] Indeed, it is not precisely clear what sort of distinction the name *prehistory* enforces.

Much of the time, the borderline of prehistory has not fallen between human history on one hand and the natural history of subhuman primates on the other, but has been a matter internal to the description of the affairs of fully human beings. On this side of the border are such things as money, cities, division of labor, writing, settled agriculture, metallurgy, theism, and science. On the farther side are hunting and gathering, nomadic herding, stone and bone, animism, and magic. Various writers move this or that achievement across the dividing line but in most such narrative treatments the general picture is remarkably the same. In another interpretation, perhaps more common today than a century ago, prehistory is not so much a sharp dividing line between areas of human existence as a broad province in its own right. Its borders are not sharply defined and its interior is not well mapped. On the far side of this province of prehistory, that is, beyond its further border, dwell beings not yet human: the protohuman primates whose lives blend smoothly into the generic landscapes of the Age of Mammals: volcanoes and swamps to the left, grasslands and icy alpine peaks to the right. On the near side of prehistory is the province of civilization, peopled entirely by humans, subdivided by technological criteria, and, where our archaeological record is ample, passing smoothly into the period of written records.

The continued existence over a period of 150 years of the province of prehistory distinct from both geological and histor-

Prehistory

ical time alerts us to the purpose of this designation, and the reasons why modern historians, like their nineteenth-century predecessors, are reluctant to let it drop. To abandon prehistory would be to postulate continuity between the biological descent of hominids, and the "ascent to civilization" of the abstract "mankind" of humanistic historical writing. Prehistory is a buffer zone. It is a place where merely biological hominids turned into "Men."[2] Somewhere in this indistinct topography, a great mental chasm marks us off from our biological predecessors. At some point a leap took place, a mutation, an explosion of creative power—the "discovery of mind," or the "birth of self-consciousness"—interposing a barrier between us and our previous brute, merely biological existence. This is the tale that most writing on prehistory has told from the last century to the present, even including very recent, very generous, and very readable books such as John Pfeiffer's *The Creative Explosion.*[3]

It is our mythology—our origin story. It has been many times reworked in the details, but the plot remains the same. In what follows, we shall trace some of the major modifications of this mythology, always with attention to the ways in which the archaeological record of prehistory has been interpreted to keep prehistory available as the "birthplace of mankind."

In the mid-nineteenth century, in keeping with the original practical designation of the term *prehistory*, historians marked the postulated discontinuity in mental achievement between historic and prehistoric mankind by insisting on the fundamental significance of the invention of writing and the keeping of written records. This plausible distinction was not successful in establishing the sharply defined onset of superior mental capacity. A higher and more successful barrier was then established through the inclusion, as a demarcation criterion, of the presence of art in historic cultures. "Art" meant not the decoration of quotidian instruments with pleasing designs but rather fine art, painting and sculpture. The notion appears to have been that the existence of works of art reflected a world view comparable to the modern. More importantly, it suggested a

Natural Knowledge in Preclassical Antiquity

leisure to paint and draw and sculpt, indicating a cultural elevation above the full-time preoccupation with livelihood.

The existence of fine art, and of an aesthetic sensibility and sensitivity to turn surplus energy toward spiritually edifying outlets seemed a good way to distinguish "us" from "them." This identification of high culture and the onset of modern mental capacity signalled jointly by writing systems and fine arts, particularly painting and sculpture—the bonds to classical antiquity being particularly strong here—became a guiding indicator to separate the "before" from the "now." It was a good barrier, and it seemed to work.

Civilizations seemed to have *both* writing and art. The distinction between writing and art assemblages (civilization), art assemblages (barbarism), and assemblages with "decoration alone" (savagery) was not always crudely drawn, I hasten to add. Not all who acknowledge distance between themselves and barbarians imagine that the latter howl at the moon. Tacitus, in writing his *Germania,* found much in the barbarian mores of Germany that Romans might do well to emulate. Diderot's *Supplement to Bougainville's Voyages* enthusiastically endorsed the sexual morality of the Polynesians over that of eighteenth-century Europe. Nevertheless, even these unforced acknowledgments of the nobility of savages and barbarians never assert that they are entirely like us; they only remind us that civilization is not an unmixed blessing. The later nineteenth century was happy to ratify and extend the distinction.

This barrier was toppled in its turn by the discovery in the 1870s of very ancient cave-paintings in Altamira, Spain, and, in the 1880s, in a succession of other Spanish and French caves, and of sculpted ivory statuettes of comparable antiquity from Brassempouy, France, in the 1890s. Once the initial disbelievers had reluctantly admitted that these were probably not all forgeries or improperly dated works of lesser antiquity, the guardians of European culture set about to assimilate the self-contradictory notion of a "prehistoric art." There were courageous souls who went over wholly to a new view of prehistory. As Gabriel de Morillet (1821–98) put it in 1877, "this is

Prehistory

the infancy of art, not the art of an infant." These were not just any sort of drawings or paintings or figures, but "Paleolithic Art . . . begins to be treated as it should be, as the first artistic manifestation of man in Europe . . . the first chapter in a long history of artistic achievement."[4] The existence of fine art in a prehistoric setting meant that if some fatal transformation of the human spirit had taken place, it had taken place many millennia before the onset of history, as then defined, and far before the advent of any of the other elements in the roster of civilization. Anyone who could produce such sublime work clearly belonged to history, not prehistory, as the term had been construed; this was not primitive culture, this was *civilization*. As cave after cave was explored and documented, folio after folio of beautiful drawings and, later, photos of the originals appeared. Cavalcades of deer, auroch, cattle, ibex, reindeer, mammoths, and horses arched across the ceilings and walls of ancient caves, like so many prehistoric chapels, and with the publication of these cave paintings came an entirely new appreciation of the spiritual life of prehistoric humanity.

It is here that the interesting work begins, for the interpretation of cave art has become one of the great intellectual preoccupations of the twentieth century. By now it has a gigantic literature of its own, with its own metaphysics, and its own rules, and it is well worth an examination in some detail.[5]

The cave paintings were, precisely because of their visual appeal, a serious challenge to the chronology of the evolutionary history of human culture. If painting and sculpture could flourish in the absence of money, museums, monotheistic religion, cities, and writing, then none of these things were necessarily emblematic of higher mental capacity. This unsettling news threatened not only the established picture of how we came to be (we were hardly over the surprise of not being descended from Adam and Eve) but questioned what the history of human life might *mean* if it were not a kind of coordinate progress in all aspects of civilization at once. The rationale for much of historical writing on human cultural evolution has been to use an ascending sequence of cultural remains to tell

Natural Knowledge in Preclassical Antiquity

the story of how we got as smart as we are today. If, using the same criteria, material remains show that we have been this smart since 40,000 B.C., some other explanation must be found to infuse dramatic tension into such cultural milestones as the Golden Age of Greece.

Something, then, had to be done about the prehistoric art problem and historians went to work on it right away, lest the whole conception of human cultural history as an ascending trajectory collapse. The successful explanatory strategy, developed early in the century and modified only slightly since, has been to turn attention from the technically superior execution of the art to its meaning. The fact of its beauty, or artistic merit, or genius was soon palliated by the "discovery" of the use to which it was put. Whatever the source of this ingenious diversion of attention (it was probably the new anthropology, which dictated that everything in a culture, even fine art, has a hidden function), at a stroke it turned cave art from the leisured philosophy of superior beings into an element of the toolbox of savages. What separates the abstract painting of the chimp from that of the $200,000-a-work product of a great culture-hero are not the subtleties of execution, but the intent and the context; what separates cave art from modern art is not its appearance, which is superficially similar, but its cultural function.

The functional significance proposed for cave art was its employment in ritual. Yes, nodded savants all over Europe (with evident relief), these paintings have *ritual* significance. At a time when advanced techniques of European exploration and colonial domination brought Europeans into sustained contact with "natives," a concept of the *primitive mentality* held sway.[6] The primitive mind, it was argued, while not precisely like that of a child, is perhaps in the adolescence of the human spirit. It doesn't think coolly and rationally about its problems; it is somewhat demon ridden, and inclined to overhasty reliance on petitionary and intercessory prayer to accomplish what rational calculation does better. In the context of a hunting society preparing for the hunt, the primitive mind goes and paints a picture of the prey animal, and then enacts rituals

Prehistory

to begin the game, and acts out the hunt and the kill. Thus the knowing nods—these cave people were primitives, and they had lots of rituals, particularly the hunting ritual. This art is not primitive in its execution but is so in its intent, and in its cultural context.

Prehistoric art was here deftly severed from the leisured, contemplative, and passionate expression of cultural energy and spiritual longing, which we all know is its true significance in the modern West, and turned into an instrumental adjunct to a hot lunch. The Paleolithic cave is no Sistine Chapel, no Paleolithic Louvre, but rather a troglodytic voodoo hall, or at best the ancient equivalent of the trophy room.

This explanation had the added strength that it gave a clue to the mystery of why all this excellent cave art is way back in inaccessible, frightening, uncomfortable, wet, dark caves where nobody can see it. The suggestion that it was meant not to be seen by *just anyone* seemed in marked consonance with the habits of contemporary "primitives," who tended to conduct their ritual out of sight of those whose presence might disturb or defile them, especially the women. These paintings were, then, the very stuff of secret rituals, which, as anthropology documents, are generally conducted by groups of the same sex. Since prey animals are portrayed, and since only men hunt, the prehistoric painting and sculpture was identified as men's business. If the women saw it, they might anger the animal spirits or the gods. (How it was divined that only men hunted or did the painting is beyond me.)

This all suggests that the mind of the primitive has no linear sequence but is a melange of custom, habit, and hearsay, that primitives were fearful and cowering, leading the nasty, brutish, and short existence dictated by the lack of the civilized comforts a rational being would have quickly acquired. Interpreted this way, the beauty of the art could rest unthreatened and unthreatening in its stone age setting because it was painted, in secret, by cold, shivering, hungry primitives who, whatever their artistic genius, thought it could fill their wretched bellies.

Natural Knowledge in Preclassical Antiquity

To make this picture even darker, the historical interpretation of cave paintings soon incorporated the persistent fantasy that "early man," who painted these marvels, had a mystical notion of the relationship between hunting and sex, and that every time he saw something pointed and hard go into something rounded and soft, his fancy turned to love. This was supposed to be so even when the pointed hard thing had six barbs on it, and the soft thing was a raging bull.[7]

Not quite supplanting this interpretive foray was the Freudian version that, by finding the same linkage of sex and violence in all of us, demarcated the savage (prehistoric or contemporary) from the civilized being by asking not whether a given society could write and paint but whether it repressed the desire of every male to kill what and whom he could, and mount whom he would. Freud defined civilization, and by association, the impulse to create art, as a sort of evolutionary booby prize we get when we can't have all the sex and violence that we want. This robs civilization of many of its most pleasing associations but it reinforces the concepts of "savagery" of the "primitive mentality" every bit as much as the less sophisticated attributions that preceded his version of the hunting-and-sex-ritual-and-magic theory.

As the twentieth century has worn its bloody way to the present, the scholarly impetus to separate the savage from the civilized through comparison of belief systems and levels of spiritual attainment has now and again lapsed into embarrassed silence. Eric Remarque's *Im Westen Nichts Neues*, with its record of the graduates of the gymnasium meeting the graduates of the lycée and of the public school in No Man's Land, in order to exchange blows with sharpened trenching shovels and shoot each other in the stomach with flare pistols, documents the exhaustion in the 1920s of any pretense that Western Civilization was an outcome of evolutionary moral development of the species. The many attempts in the 1930s to repair the distinction between savage and civilized were refuted even more horrifically by what followed between 1939 and 1945.

Prehistory

One might suppose the game would be given up altogether at this point. That it was not reveals the power of the evolutionary metaphor, and its importance in the structure of modern thought. In the period after the Second World War, more books appeared on cave art than ever before. These approached the material in a somewhat different way, which reflected a change in the concept of human evolutionary development. With the impetus to infuse human evolution with moral meaning at least temporarily out of the way, interpretation of cave art tended to focus on a new objective: refixing the boundary of "us" and "them" just *before* the period of the cave paintings. Something like this had been tried in the context of species evolution by claiming that cave art appears just as Neanderthals disappear, and therefore that the cave art came conjointly with a respeciation of the genus *Homo* in the direction of *Homo sapiens*. This theory, of which there are still audible echoes (see, for instance, the Cave Gothic novels of Jean Auel, especially the *Clan of the Cave Bear*), perished scientifically with the admission that Neanderthals were not a different species after all, and attempts to make them stooped, gnarled, chinless, and musclebound (the familiar representation) were ultimately based on the mischance that the first Neanderthal skeleton unearthed was that of an old man with advanced osteoarthritis. Archaeology has since further complicated the picture by digging up fossil remains of *Homo sapiens* that are clearly older than most Neanderthal skeletons: enter "archaic *Homo sapiens*." The physical evolutionary picture of human development no longer seemed capable of supporting a radical evolutionary development which might have left decisive physical evidence in the "recent" past—40,000 years ago.

The publication of researches by investigators in East Africa, who in a single generation have pushed hominid evolution back more than three million years, have made such a linear evolutionary scheme yet more difficult to accept or maintain. Even more compelling has been the increasing amount of information on the life and times of that member of our genus known as *Homo erectus*, who occupied much of

Natural Knowledge in Preclassical Antiquity

Asia and Africa for more than a million years—between about 1.5 million years ago and about 400,000 years ago. Studies of endocasts (casts molded in the inside of the skulls) of *Homo erectus* show a degree of brain lateralization, or asymmetry of the hemispheres, similar to that of modern humans and suggesting the capacity for language. This challenged a favored placement of "the barrier" at the emergence of *Homo sapiens sapiens* about 100,000 years ago, on the grounds that they were the first hominids who could talk. That this "language barrier" argument is not entirely dead can be seen in the work of paleontologists Niles Eldredge and Ian Tattersall in *The Myths of Human Evolution* (1982). They still accept the argument of George Sacher that language appeared instantaneously when the human brain reached a "critical size": "Sacher's scenario . . . helps explain, for example, how some of the very greatest artistic achievement of mankind—the cave paintings of Lascaux and Altamira, for instance—were made so soon after the first appearance of securely dated modern man in Europe." What is mainly evident here is the fervent desire for a wall, for they also confess that "Sacher's scenario [is] impossible to prove."[8]

Our archaeological encounters with new predecessors who could walk and talk, who lived in houses and family groups, who made tools, and who cooked food more than a million years before we appeared as a species has had the interesting effect of making the cave painters seem much closer to us in comparison, and has increased the pressure to bind them to us while repairing the barrier which keeps "them"—our earlier ancestors—back. If the language barrier is doomed to fail (since the cranial capacities of *Homo erectus* overlap those of *Homo sapiens* by more than 200 cubic centimeters, the critical mass had to have been achieved more than a million years ago), then some other criterion has to be found to replace those that have slipped away. Using biological differentiation to define evolutionary advance in mental development, and then using advanced mental development as an explanation of cave art as a barrier between us (now including the cave artists) and our predecessors is not after all scientifically founded on morphological evidence bearing on cranial capacity, brain lateralization, or cerebral architecture. Such a barrier does not work. This lack of a meaningful explanation has challenged scholars to develop new theories or modify older ones to preserve the notion of progressive advance and recent emergence of "mankind."

Within the confines of true (i.e., biological) evolutionary theory there is currently a debate over the mode and tempo of evolutionary change. The classical neo-Darwinian synthesis of evolutionary theory postulates a constant variation of species through time by the mechanism of natural selection. It is a uniformitarian theory, which means that it postulates that the world has always changed at about the rate, intensity, and by the same means we see in effect all around us today.

Contrasted with this neo-Darwinian synthesis is the theory of punctuated equilibria, which argues that the fossil record shows long periods of biological stability in many species, punctuated by rapid, radical, intense episodes of extinction and respeciation, resolved in a new and different equilibrium. It is a catastrophic theory, which means that it sees marked evidence of occasional significant variation in the rate, intensity, and even the means of evolutionary change.

It has not escaped the attention of those whose reading is up to date in archaeology and physical anthropology that this model fits the record of human evolution quite well. Eldredge and Tattersall's book was written in defense of the application of the theory of punctuated equilibria to hominid evolution. To make the case, they refer to the evident stability of *Homo erectus* and his tools, from 1.5 million years ago to about 400,000 years ago. Then there is a period of wide variation in human form from 400,000 to about 35,000 years ago in which a broad spectrum of human skull morphology is evident, characterized as archaic *Homo sapiens, Homo sapiens neanderthalensis,* and by a number of skulls which do not fit any of the

Natural Knowledge in Preclassical Antiquity

morphological categories. After 35,000 years ago there is again stability, marked by the universality of the lighter bones and skulls of *Homo sapiens sapiens*.[9] This renewed attempt at a plausible barrier abandons the picture of evolution as a progressive and gradual development, and substitutes long periods of static continuity very occasionally interrupted by episodes of rapid change. From the standpoint of our examination of the relentless urge to separate modern mankind from its evolutionary history as a hominid species, the outcome is the same: we are in this construction plausibly seen as the products of a radical break.

I have alluded already to a striking recent attempt to capitalize on this picture of long-term stability and rapid and even catastrophic evolutionary development: John Pfeiffer's *The Creative Explosion: An Inquiry into the Origins of Art and Religion* (1982). Pfeiffer sees cave art as one expression of a number of simultaneous developments: large-scale social organization, differentiation of social status, specialization of labor (including shamanism and knowing as a specialty), and emergence of religion, art, and personal ornamentation, all of which he dates from about 40,000 B.C. Pfeiffer sees this creative explosion into art as "one of a number of developments marking the efforts of people like ourselves to adapt to the release of formidable evolutionary forces, forces which brought about changes in a lifestyle that had endured unchanging for millenniums." But why, he asks, was there art? "What was on the artists' minds? Why did they go underground—and always why then, why after more than a million years of *homo* and no enduring traces of art, starting a mere 30,000 or so years ago, with a notable spurt 10,000 years later?"[10] Art is a response, says Pfeiffer, to evolutionary stress, based on population pressure and the associated responses—wider and more intense exploitation of the "wilderness" for food, changes in tool technology, and wider migration and expanding exchange networks, all of which required some kind of extrasomatic, mnemonic recording device to function in a ritual context and to manage the information explosion.

Prehistory

The accompanying chart (see p. 16) from *Creative Explosion* is an admirable summary of the thesis: art moves from archaic to primitive to mature, as tools ascend step-wise to greater fineness and sophistication, with a rhythm apparently governed by glacial maxima and minima and associated pressures, with the sharpest break coming at the postulated Glacial Peak of 20,000 years ago, when both tool culture and cave art experienced a spurt of growth and diversification. All of this takes place within the horizon of dominance of *Homo sapiens sapiens,* which stabilized (as Eldredge and Tattersall put it) as a species about the time Pfeiffer's chart begins.

Before accepting this attractive and plausible chronology, let us consider the implement in figure 1.1.

It is the most abundant and important tool in the history of the genus *Homo.* It was in continuous use for more than a million years. But we don't know what it is, or what it was used for. It is commonly referred to in English as an Acheulian hand-axe, a reference to the locale where it was found first (St. Acheuil, France) and to its size and shape—rather like an axe-head and finished with a sharp edge all the way around, typically about as big as the palm of a large hand.[11] Microscopic analyses of the edges of hundreds of such "axes" reveal no evidence that they were hafted (bound or wedged into handles). They were therefore hand-held or hand-hefted (depending on how they were used). This makes it extremely unlikely that they were used as axes. For one thing, their sharp edge all round meant that the first hard strike into a resistant object would leave a severe wound; Newton's third law applies in the Paleolithic world as well as our own.

The ubiquity of this stone tool and its long survival suggest that it was multifunctional, the Swiss Army knife or Collins machete of the Paleolithic world. One could kill an animal with it, gut it out, carve it, and slice it, work wood and bone with it, and so on. Quite recently Eileen O'Brien has suggested that these were projectile weapons.[12] She noted that we find them in the sites of ancient watercourses, indicating that the ones we find were lost in the hunt—an interpretation that gains

Natural Knowledge in Preclassical Antiquity

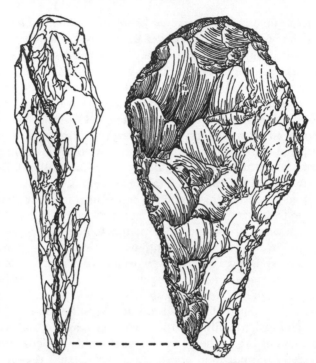

Figure 1.1 Acheulian hand axe, the principal tool employed by the genus *Homo* for more than a million years. Its precise function is unknown. Redrawn after A. Rust, *Die Höhenfunde von Jabrud (Syrien)* (Neumünster, 1950).

some credence from the number found embedded point downward. Their size, symmetry, and sharpness indicate that making them required sufficient time, energy, and care that one would not toss them away lightly. To test her hypothesis, Dr. O'Brien, then at the University of Georgia, induced some of the track team's discus throwers to try throwing it. After forty tries one of the trained discus throwers could consistently put the projectile in a circle six feet in diameter from 100 feet away. When one imagines the skill level attained by beings who did this for livelihood over a lifetime and hunted in groups, one

Prehistory

could see that the hunting efficiency of even a small group could be very great indeed.

This ingenious and empirically well-founded hypothesis is not, however, my point. My point is that we do not know unequivocally what this tool was used for, and this degree of uncertainty is typical of our knowledge of Paleolithic cultures until quite recently.

The historical scenarios of theories of "creative explosion" and biological emergence of true *sapiens* in very recent time depend on the hypothesis that diverse and finely worked tools in stone and bone, cave art, items of personal adornment, and musical instruments and other cultural artifacts above the level of subsistence tools appear rather suddenly and nearly simultaneously about 40,000 years ago, about the time that hominid skull morphology settled into the familiar *Homo sapiens* character. It is by no means a silly hypothesis or a misreading of available evidence to suggest that a cultural leap took place in tandem with a respeciation of the genus, if this latter in fact took place.

The notion of an evolutionary emergence of ever more finely worked tools has been a standard part of archaeological theory for more than a hundred years, beginning with the emergence of "blade" tools about 40,000 years ago, and "ascending" through a sequence of types. One such sequence, taken from a book published since 1980, appears in figure 1.2.

These sequences unquestionably exist, although the record is one of emergence of new styles, not the disappearance of older ones—and most of the more "advanced" styles are only a small percentage of the tools exhumed at any level in the sequence. Nevertheless, the emergent character of these tool shapes is regularly offered as evolutionary evidence contrasting invidiously with the Neanderthal tools of the period 50,000–120,000 B.C., which have the cultural designation of "Mousterian." Mousterian tools, which themselves show a great deal of variety and specialization of function, are contrasted with the more meager remains of the Acheulian handaxe cultures which precede them. So the archaeological record

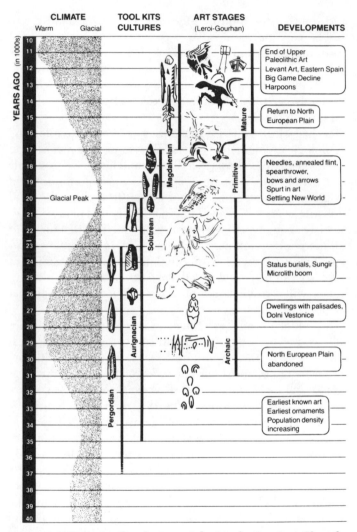

Figure 1.2 Major art developments during the Ice Age. Chart from *The Creative Explosion* by John Pfeiffer. Copyright © 1983 by John Pfeiffer. Reprinted by permission of HarperCollins publishers.

Prehistory

is still largely interpreted in a biological, evolutionary, species-specific way—with *Homo erectus* tools, *Homo sapiens neanderthalensis* tools, and *Homo sapiens sapiens* tools succeeding each other in time, rather clearly distinguished by their increasing variety, specialization of function, and fineness of construction.

This periodization is and has been enormously influential in archaeology and geology. The geological chronology of the European Pleistocene was until recently based on the use of human tools as index fossils, an assumption that they showed a clear evolution through time, and could be correlated to create time-planes with strict stratigraphic meaning. Thus the European Pleistocene from 600,000 B.P. down to 10,000 B.P. (before the present era) was classified such that the appearance of a certain form of tool in a stratigraphic layer could be taken as independent evidence for the age of the layer. More recent geological work has not only shown the dangers of this scheme, but indicated that there is no absolute ascending sequence of tools of greater perfection in Europe—that tools of greater and lesser refinement alternate in the record over hundreds of thousands of years, and that it is possible that the character of the tools is an index not of the sophistication of their makers, but of the kinds of environments in which they were used.[13]

That there is a good deal of wishful thinking in schemes of linear ascent is quite evident, and nothing in the recent history of archaeology has brought this home more forcefully than the technique of "microwear analysis" of stone tools. These analyses, conducted by microscopic examination of the working surfaces of stone tools, are based on the hypothesis that tools used for specific purposes and used on specific materials (meat, bone, wood, stone, vegetable matter) acquire a characteristic polish, specific to that material or use. This polish is different from the larger striations and use-damage visible with low-powered microscopes and is difficult to detect.

The technique of microwear analysis stems from the early studies of the Soviet prehistorian S. A. Semenov, whose *Prehistoric Technology* was published in English translation in

1964.[14] It was coolly received, both because of the difficulty of replicating his results (early imitators used microscopes of insufficient resolving power) and perhaps because Semenov's own conclusions seemed merely to broadly reconfirm the hypothesis of successive cultural complexity, in close conformity with the ideological necessities of dialectical materialism.

Analysis of stone tools by Lawrence Keeley in the 1970s, however, not only contradicted Semenov's cultural evolutionist results, with their linear ascent of sophistication in manufactures, but threw the entire evolutionist scheme into question.[15] Keeley made a number of technical refinements in the method of microwear analysis and set to work on some stone tools from Hoxne, England, representative of a cultural assemblage known as Clactonian, associated with an interglacial phase of about 300,000 years ago. His analyses of the varieties of polish, the abrasions of the surfaces of stone tools by the materials on which they work, on different tools indicates that in addition to cutting meat and scraping hides (each of which produces a diagnostic polish), these Clactonian tools were used for woodworking—scraping, shaping, and boring—and also used for boring bone, and slicing or cutting plant materials other than wood.

Keeley's results take on a special significance because, in anticipation of criticism of the technique's adequacy and accuracy, he submitted himself to a blind test. He had a colleague make a series of stone tools, use them in a variety of settings on different materials, and then submit them to Keeley for microwear analysis. Keeley was able to make thirty-six correct identifications out of a possible forty-eight.[16] The test led to a further refinement of his techniques of analysis which pushed his success rate (on real tools) to between 70 and 80 percent accuracy.

Two important generalizations seem possible here, both of which cast doubt on sweeping hypotheses of cultural emergence linked with biological evolution in recent time. The first is that the analysis of the Hoxne tools indicates that most of the stone tools, other than those used to cut meat or scrape

Prehistory

hide, were tools to make tools, and that with very few exceptions (one of these being a wooden spearpoint and a shaped club from Hoxne) these bone and wooden tools have not survived. The second and related point is that cultural reconstructions based solely on surviving stone tools must, therefore, give a profoundly impoverished view of the cultural achievement of an early period.

Under these circumstances, we may ask whether or not resolute schemata of cultural evolution based on increasing profusion and fineness of tools do not confuse the extent to which cultural elements have survived in the archaeological record with the level of sophistication achieved. It is all too easy to confuse the gradient of preservation with the level of sophistication. Not surprisingly, those cultures close to us in time, those cultures since the last major glacial advance, and those cultures which regularly inhabited caves have left a more ample record of artifacts, more suggestive of complex social existence. Cave paintings do survive, after all, because they are in caves. For all we know there were beautifully decorated murals painted on mammoth hides a quarter of a million years ago; the Paleolithic equivalent of *War and Peace* could have been carried about for millennia, and we would never know about it, if it were made of perishable substance. While we have many durable implements from the European Middle Ages, the scant remains of cloth and leather used by the common folk are those preserved in bog burials or from frozen graves in Greenland, where the Norse settlers of the twelfth and thirteenth centuries put their dead to rest in permafrost. A culture as abundantly supplemented with artifacts as any nonmetal, nonpottery, Neolithic culture in the world could have existed for two thousand centuries and we would know nothing about it.

It is problematic on this time scale whether our own culture will leave anything more permanent. If the ice returns, and the scientific odds are that it will, then we can expect the surface remains in North America above a line from Portland, Oregon, to Hackensack, New Jersey, to be pulverized. Any-

Natural Knowledge in Preclassical Antiquity

thing that won't freeze and crush will rot and crumble. This somewhat gloomy prediction (depending on your taste; a new broom, even a glacial one, sweeps clean) is only an illustration of the wrongheadedness of assuming that the implements that last 250,000 years are a good index of the variety of implements a culture possessed.

On this scale, archaeologists of the future will find evidence of a high culture with imposing structures around 1300 A.D., which thereafter went into unrelieved decline—the stones ever less well carved, the stone buildings smaller, until the culture died out altogether somewhere after 2000 A.D. Using the canons on which cave art is ranked in evolutionary sequence, Henry Moore and his imitators, should they survive long burial, will compare unfavorably with nineteenth-century copies of Greek originals, and will suggest, if anything, the ubiquitous "Venus figurines" of the Paleolithic. Without protection a stone inscription can last 10,000 years, a clay tablet nearly as long, a parchment a few centuries. A book printed on low acid paper and kept in a controlled climate might last as long. The shelf life of protected metallic or plastic recording tape is unknown, but left unprotected could be as short as a few years. The more advanced a culture's material science becomes, the less durable it is—fineness and delicacy are the first to perish.

There are two equations at work here. One is the equation, expressed as an evolutionary series, of the archaeological remains of any period of hominid development with the level and quality of culture achieved. This, as I have already noted, makes most of its gains by discounting organic decay of all material not made of stone, and ignoring the fact that all the "art" from which the cultural ascendancy of the later periods is assessed is either painted on the stone walls of caves, made of stone, or of precisely those materials (bone and wood, skin and fabric) which have organically perished from the remains of the earlier periods. Note that this equation cannot be defended by the empirical claim that we stick to the evidence, and that we can argue only from what survives. Keeley's mi-

Prehistory

crowear analysis of stone tools more than a quarter of a million
years old documents their use on hide, bone, wood, and plant
materials, so we know that we have only the stone tools, and
that since these were largely tools to make tools, we lack most
of the tools and implements made. We have then no evidence
of an ascent or explosion or any other sort of directional sequence that cannot be dismissed as a spurious invention based
on the misinterpretation that the archaeological record is essentially complete.

The second equation equates the level of intellectual and
spiritual culture to the complexity and character of the surviving material remains—to the extent of postulating the
arrival of complex social organization, religious belief and
ritual, and image making and personal adornment simultaneously, very recently, and, not incidentally, in Europe, at the
archaeological levels at which the gradient of preservation begins to yield large amounts of material remains not made of
stone. Given our first conclusion, that the archaeological record
of human cultural remains is radically incomplete, and the
more incomplete the farther back one goes, we would have to
ascribe all the achievements of the cultural explosion to all
hominid groups in the last quarter of a million years, and perhaps earlier.

It seems to me that the evidence demands we do this.
But the evidence does not demand that we accept the fundamental equation of material with spiritual culture, and particularly the equation of the latter with art and religion, which
provide the impetus for it in the first place. The entire
complex of evolutionary sequences, of ages and epochs,
however sophisticated, serves in every period of the intellectual history of the modern world to maintain a fixed vision
of the ascent from savagery to barbarism to civilization in
measurable steps. It is not a mean-spirited vision, but its
purpose is to maintain the myth that we are somehow special
and different or even new, in some significant sense, from
our nearest hominid ancestors. It maintains this myth by
erecting new barriers and new means of exclusion as old

Natural Knowledge in Preclassical Antiquity

ones fall—writing, painting, language, and cranial capacity
have all had their days as indexes of "us." The latest versions
reshuffle this material, but deal us a myth of emergence
which differs in few details from anything that the cultural
materialists of the middle of the last century (Morgan, Spen-
cer, Engels) would have been happy to embrace.

History as reporting, critical history, as even the Hellenic
Greeks of the fourth century B.C. practiced it, is corrosive. Its
canons of truth are not concerned with what comforts, what
elevates, or what gives hope. It lives by attaching itself to
myths and growing at their expense, by dissolving their embrace of a vision of "us" as better than "them." The recent
history of prehistory and the march of archaeology as a discipline constantly provide new materials for the reworking of
these myths of ascent, but the overwhelming tendency of archaeology (including hominid paleontology) has been to make
such myths increasingly difficult to substantiate, and increasingly pointless.

Oswald Spengler was a good second-rate thinker who had
the knack for distilling the conclusions of the previous generation or two of first-rate thinkers into readable form. He noted
that the scheme of the history of European civilization as ancient, medieval, and modern was "hopelessly jejune," and
pointed out the absurdity, the wrong-end-of-the-telescope
character of seeing 4,000 years of Chinese history through a
lens which swelled up to gigantic proportions European history since 1500 A.D. It is clear that the scheme "Paleolithic,
Neolithic, metallic" is just as hopelessly jejune, and sees the
last half million years, and perhaps the last 3 or 4 million years
of hominid existence through the wrong end of the telescope
of European history in the last 30,000 years. There are more
zeroes after the numbers in the latter comparison, but the dis-
tortion ratio is the same.

Spengler went on to say that the "Western consciousness"
seems to find itself urged to predicate a sort of finality inherent
in its own appearance.[17] This is captured in our use of the
word *modern*—a designation sliding ever forward as time rolls

on. Yet Spengler also points out that at the very time the designation *prehistory* was first fashioned, the 1820s, the philosopher Arthur Schopenhauer declared that our problems are not the problems of "abstract man," but of actual man on the earth in historical time.

Prehistory is the last stamping ground of abstract man, and exists within that frame. If we are entering a postmodern age, accompanied by declarations of the "death of [the concept of] man" and the "end of [linear progressive] history," we might expect that one of the indicators of its arrival will be the annexation of the province of prehistory into history, and both of these into planetary and ecological, if not geological time. In the meantime one may wonder what new definition prehistory will take if it can indeed survive the demise of most recent mythological recasting of the Ascent of Man.

2

Egyptian Fractions

Some of the grotesque miscalculations and misappraisals that have been made in comparing the working efficiency of past ages with the present . . . have committed the blunder of confusing the increased load of equipment and the increased expenditure of energy with the quantity of effective work done . . . The fact is that an elaborate mechanical organization is often a temporary and expensive substitute for an effective social organization. *Lewis Mumford*

Wisdom of the Egyptians

The pyramids had been up for a thousand years when the whole corpus of Greek literature was still little clay tablets that said things like "2 tripods, 1 four-handed goblet" (Pylos Tablet Ta 64) and "163 rams, 54 she-goats, 44 ewes" (Pylos Tablet Cn 131) in Mycenaean Linear B.[1] It was another thousand years before Herodotus got to Egypt, and we should not be surprised that he was stunned by the antiquity and size of what he saw.

Thus began the tradition (mightily encouraged by the Egyptians) that all the valuable learning of the world could be traced back to the valley of the Nile. Thales of Miletus (whom we shall meet later in another context), who was honored by the Greeks as their first geometer, was not supposed by them to have invented geometry, but to have gone to Egypt to learn it. When Democritus sought to advertise his skill in geometric proofs, he boasted that even the Egyptians could not do it any better. When Johannes Kepler, in 1618 A.D., discovered his third law of planetary motion (that the squares of the times of revolution of all the planets are proportional to the cubes of

Egyptian Fractions

their distances from the sun), he wrote: "I have robbed the golden vessels of the Egyptians to make out of them a tabernacle for my God, far from the frontiers of Egypt."[2]

The tradition of Egyptian wisdom has been remarkably persistent, with cultural roots in Hebrew, Greek, and Latin history. It has enjoyed a renaissance in every century of the modern age—the seventeenth, eighteenth, nineteenth, and twentieth. The most recent perigee was attained in the Tutankhamon craze of the 1970s, the only time in memory when scalpers made money selling tickets to an art exhibit. At about the same time there was a resurgence of the "cult of the pyramids" and of "pyramid power." The latter proceeded on the assumption that the pyramidal form has an ability to focus cosmic energy (and that the Egyptians, of course, had known this) and can keep milk fresh without refrigeration, keep razor blades sharp, improve sleep, retard aging, and accelerate the healing of wounds. After spoiling a lot of milk, causing many shaving injuries, and reaping millions in the sale of pyramidal frames of various sizes, the movement subsided.

This subsidence can only be temporary, for of all the monumental architecture of the Ancient Egyptians that ignites the imagination of visitors to the Nile it is the Great Pyramids that are emblematic of the civilization. Until quite recently they were the largest structures ever built, and the aura of mystery around them is enhanced by the "paradox" that they are among the oldest. The Great Seal of the United States carries the image of a pyramid with an all-seeing eye at the top; the secular secret orders of the eighteenth century—Masons, Illuminati, Rosicrucians—in whose ranks many of the founding fathers were numbered, all had special places in their arcana for Egypt, and for pyramid lore.

Any imagined slight against the Egyptians calls forth a retort about the pyramids. Robert Benchley once wrote an irate letter (I am told) to the Metropolitan Museum in New York after seeing an exhibit label in the Egyptian collection that directed the viewer's eye to "a remarkably life-like picture of a goose." "Might we not suppose," said Benchley, "that a civilization

Natural Knowledge in Preclassical Antiquity

capable of erecting the Great Pyramids could also draw a goose?"

A corollary to the postulate of Egyptian wisdom has always been that it was esoteric, arcane, hidden. This assumption is not too difficult to concede but even the most enthusiastic supporters of the Egyptians have been more or less driven to the postulation of hidden wisdom by the yawning gap between the primitive character of their technics and the great beauty, size, and precision of their architectural executions.

For the pyramids are not just ancient and huge, but beautifully exact. So much so that Charles Piazzi Smyth, Astronomer Royal of Scotland and a professor at the University of Edinburgh, wrote a treatise in 1864 (*Our Inheritance in the Great Pyramid*) in which he carried out a number of numerical manipulations with the dimensions of the pyramid of Khufu to prove that they contained hidden correspondences with the dimensions of the earth, and were inspired by God, and left as a vehicle of prophecy.[3] Martin Gardner, in *Fads and Fallacies in the Name of Science*, said of his work: "few books illustrate so beautifully the ease with which an intelligent man, passionately convinced of a theory, can manipulate his subject matter in such a way as to make it conform to previously held opinions."[4]

This is perhaps unfair, for the exactitude through which Smyth dreamt his dreams is actually there. Sir William Flinders Petrie, the greatest of all the British Egyptologists, got his start in the business when sent by his father, a disciple of Piazzi Smyth, to check the measurements of the Great Pyramid. He found that the error of length of the four 755-foot sides is less than 1 part in 4,000; the error of squareness is little more than 1 foot; the error of levelling is less than 5 inches a side, and not 12 inches in total, and for any 50-foot length it is less than ⅕ of an inch. The four corners of the pyramid are exactly ordered to the points of the compass.[5] And so on. The apparent mystery is that this was accomplished in a civilization which has left record of surveying tools no more complicated than a forked staff, a plumb-bob, and a knotted rope. When I say left record I do not mean that we have inferred this from scanty remains: the Egyptians painted elaborate and detailed tomb frescoes of measurements being carried out in this manner.

This primitive apparatus has spurred the assertion that there must be much that the Egyptians knew (in order to accomplish these marvels) that is now hidden from us—a marginally controversial claim, to be sure, for any civilization known archaeologically. A separate but similar strategy pursued to remedy this perceived asymmetry between the methods and accomplishments of the Egyptians has been the claim that they had external assistance, whether in the form of divine aid, as in Piazzi Smyth, or with help from people in flying saucers, as broached by Eric van Daniken (*Chariots of the Gods*). This species of assertion reduces to the following: that given their primitive technics, the Egyptians had either great hidden wisdom or wise helpers, for they could not have done what they did with what we have determined they knew. In Piazzi Smyth's case, Gardner has already spoken. In van Daniken's, we might wonder why the flying saucer people were so free with their knowledge of civil engineering and so secretive about dentistry, given the number of royal mummies with abscessed and rotting teeth.

More recently and more interestingly, a polymer chemist named Joseph Davidovits has contributed to this speculative genre and argued that, rather than quarrying limestone blocks, barging them down the Nile, hauling them on sledges to the site, and dragging them up the sides of the pyramid, the Egyptians employed a lost technology of polymer chemistry to mold a slurry of crushed limestone with a mineral binder into beautifully uniform blocks at the building site, to make the finishing casing stones of the pyramids.[6] This lost technology would serve to explain the virtual identity of each casing stone with all the others, and would do away with what one might call the "ant-colony" theory of pyramid construction. Behind the sophistication of his argument lies a familiarly resolute unwillingness to believe that anyone who could do something as grand as to build these Great Pyramids would do it in such a

Natural Knowledge in Preclassical Antiquity

wasteful, brute-force fashion, which leads once again to the hypothesis of hidden wisdom.

To the extent that Egyptian wisdom is reflected in its most vaunted form in exact measurement in the execution of great structures, a good deal of attention has been focussed on Egyptian mathematics. Here claims of hidden wisdom have also flourished, for the mathematical tools which the Egyptians left behind are very much of the forked-stick and plumb-bob order, and this has proved a disappointment to some and a great puzzle to many others. It is an intriguing problem. If there is any hidden wisdom in Egypt we should expect to find it here, for in our experience, the core of any body of science is its mathematics.

Egyptian Mathematics

The best book on Egyptian mathematics in English is Richard J. Gillings's *Mathematics in the Time of the Pharaohs* (1972), which not only gives a clear exposition of Egyptian numerical techniques, but also provides a cogent history of attempts to match the supposed primitiveness of the mathematics to the much grander reputation of the civilization.[7] Gillings points out that, as in the case with monumental architecture, Egyptian mathematics reached its highest level in the Old Kingdom (around 2500 B.C.) and stayed there for the next three millennia, a conclusion well borne out by the documents, albeit at variance with our notion of scientific progress.

While we have records containing calculations and complicated apportionments of goods and services, land surveys, and lists of duties owed and collected, our knowledge of the operations whereby these sums and results were obtained are drawn from a very few documents: the Rhind Mathematical Papyrus, the Moscow Papyrus, the Egyptian Mathematical Leather Roll, the Kahun Papyri, and the Reisner Papyri. Gillings's conclusion, after a painstaking dissection and analysis of these documents and a review of a half century of devoted scholarship, is that "it will surely come as a great surprise to the readers of this history to find that whatever great heights the

Egyptian Fractions

ancient Egyptians may have achieved scientifically, their mathematics was based on two very elementary concepts. The first was their complete knowledge of the *twice-times table,* and the second, their ability to find *two-thirds of any number,* whether integral or fractional. Upon these two very simple foundations, the whole structure of Egyptian mathematics was erected."[8]

Employing simple techniques with the aid of sets of tables—twice times tables, two-thirds tables, tables of "auxiliary" or "reference" numbers (least common denominators), and tables as well of the oddest aspect of the Egyptian mathematical apparatus, "the unit fractions"—Egyptian scribes performed all their work of measurement and apportionment throughout the history of the civilization.

The treatment of fractions is especially interesting. Egyptian mathematics, like our own, was decimal: based on tens, hundreds, thousands. But when we come to fractions of whole numbers, the Egyptians used an astonishing system. With the sole exception of the fraction $\frac{2}{3}$, all Egyptian fractions are unit fractions; they have a numerator of "one."

To show how this worked, let us divide 9 things into 10 equal portions, comparing our method and the Egyptian method. It is an Egyptian sort of problem: how to divide equably nine jugs of beer or nine loaves of bread among ten thirsty and hungry workmen (this is, in fact, problem 6 of the Rhind Mathematical Papyrus). We would say, "Each man gets $\frac{9}{10}$ of a jug." But in Egyptian mathematics there is no such number as nine tenths and the Egyptian system, carried out with the aid of unit fraction tables, provides that each man shall receive $\frac{2}{3}$ of a jug + $\frac{1}{5}$ of a jug + $\frac{1}{30}$ of a jug. We would never leave things there—we would find the common denominator (30), carry out the necessary operations in common units ($\frac{20}{30}$ + $\frac{6}{30}$ + $\frac{1}{30}$ = $\frac{27}{30}$, reducible to $\frac{9}{10}$), and arrive at our own answer.

Of course, given the aim of most Egyptian mathematics, which is measurement and apportionment of quantities of food, drink, straw, building stone, or manpower requisite to

Natural Knowledge in Preclassical Antiquity

the matter at hand, this is not so odd as it may seem. Where we are concerned with the measurement of everyday quantities of the same sort today, we employ unit fractions exclusively. In the United States liquor is still sold by the fifth, the quart[er], and the half-gallon, with decimal equivalents grudgingly and superfluously inscribed on the containers. Butter and margarine come in quarters, and cold-cut meat and cheese are priced by the quarter and half pound.

The same constancy of unit fractions underlies the bewildering array of units of measure in the English system. A barrel is 31 and ½ gallons; a rod is 16 and ½ feet; an avoirdupois ounce is 437 and ½ grains; the fluid ounce is ¹⁄₁₆ of the pint, the dram ⅛ the ounce, the scruple ⅓ the dram, and so on. The only non-unit fraction which commonly appears is ¾—as in the perch, a volume of stone masonry in a wall 16 ½ × 1 ½ × 1 feet = 24 ¾ cubic feet. In our colloquial partitioning, the half, the third, the quarter, and the fifth are the commonest fractions in everyday use.

The ubiquity of the electronic pocket calculator is quickly obliterating a more complex survival of the unit fraction in the continued or chain fraction—a method of rapidly approximating large "proper fractions." The fractional number 3,937/100,000 can also be expressed as an "integer chain fraction," in which each denominator contains a fraction, the numerator of which is 1.[9]

$$\frac{3,937}{100,000} = \cfrac{1}{25 + \cfrac{1}{2 + \cfrac{1}{2 + \cfrac{1}{787}}}}$$

While calculation with integer chain fractions requires manipulations with numerators greater than one, the purpose is the same as that of Egyptian unit fractions—to express very large fractional quantities in simple terms, and to arrive at remainderless apportionments of the divisible quantity by the most rapid route. Such integer chain fractions are already his-

Egyptian Fractions

torical curiosities in everyday life, along with such hoary accountant's tricks as "casting out nines" and "casting out elevens" to check the accurate addition of long columns of figures, but they play a role in a field of modern mathematics called Diophantine Analysis, named after a Hellenistic Egyptian mathematician, Diophantus of Alexandria.

Certainly it is the unit fraction, of all the aspects of Egyptian mathematics, which has led to the postulation of some discrepancy between the totality of mathematical methods available to the Egyptians and those which we have recovered through archaeology. In the simple example I have given, dividing 9 units into 10 equal portions, there does not seem to be much of a problem. But let us take another problem, given by Gillings, from the Reisner Papyri, which are a series of official registers from dockyard workshops dating from the reign of Sesostris I of the XII Dynasty, about 1880 B.C. The calculation in question concerns assignment of workmen to dig a foundation excavation for a temple extension, on the assumption that each workman can dig 10 cubic cubits per day.[10]

The difficulty in the problem stems from the units of linear measure employed—the palm and the cubit. If there are four fingers in a palm, and seven palms in a cubit (there are), how much time does it take to dig a hole 3 cubits and 5 palms long, 1 cubit and 2 palms wide, and 6 palms deep? The scribe is not very far into the volumetric calculation before he must add the numbers $(\frac{1}{2} + \frac{1}{7} + \frac{1}{14}) + (\frac{1}{2} + \frac{1}{4} + \frac{1}{8} + \frac{1}{28} + \frac{1}{56}) + (\frac{1}{14} + \frac{1}{28} + \frac{1}{56} + \frac{1}{192} + \frac{1}{392})$ and multiply the result by $(\frac{1}{2} + \frac{1}{4} + \frac{1}{14} + \frac{1}{28})$; this operation produces more than twenty fractions to be summed. This, as Gillings notes, leaves the scribe "in a maze of complicated multiplication of almost unmanageable fractions."[11] Gillings assumes, probably correctly, that there were approximations and shorthand devices used, and even gives a plausible example or two of how this would work.

But as the planning of the foundation digging requires at least thirty-one calculations of this magnitude, and since these calculations are part of a document with sixteen sections, con-

Natural Knowledge in Preclassical Antiquity

cerned with food to be drawn from storehouses to feed the workmen, accounts to which these draws are to be charged, calculations of the temple floor plans, walls, trenches, and corridors, amounts of stone to be quarried to build them, and myriad other matters, the question has arisen: *why did the Egyptians not invent an easier way to handle fractions,* rather than trying to shortcut the system with unofficial and merely approximate algorithms, whose existence we can know only by conjecture? This is an excellent question to ask, and it raises profound questions about the nature of Egyptian mathematics.

What *Is* a Fraction?

Numeracy, the ability to count, is independent of literacy and clearly precedes it historically; we have inscribed reindeer-scapula batons from Paleolithic sites that have markings which are unmistakeably tallies of some sort. Moreover, as Karl Menninger's *Number Words and Number Symbols* (1958) reveals, there are sufficient different systems of numeration and different techniques of performing simple calculations in the historical record to argue that there are a number of different natural mathematical languages, just as there are different natural languages of the verbal sort.[12] Once we view mathematical languages in this way, we should study them via a kind of comparative linguistics.

In so doing we might wish to recall that one of the great milestones in the history of comparative linguistics is the work of the anthropologist Franz Boas, who argued persuasively that much difficulty in understanding the sound systems of new languages, as anthropologists tried to record them, came from the assumption that the sound system of any language could be mapped with reference to the common sound units, or phonemes, of modern European languages.[13] European anthropologists had interpreted their inability to map consistently a given language into Roman alphabetic transcription as an inability of the native speakers definitively to fix the sound boundaries of their speech units. Much theorizing about primitive sound systems eventuated before Boas pointed out that there was no reason whatever to take the transcription system common to European languages, and conformable to its phonemics, as a privileged standpoint from which to map other and unrelated languages, and that therefore linguistics must be descriptive before it could be comparative. In other words, it was not that the speakers could not speak the differences, but that the anthropologists could not hear them.

Just as sounds do not map easily and without remainder across linguistic boundaries, neither do concepts. It is quite unreasonable to demand translation without remainder—how could it be so? With true synonymy so rare within a given language, why should the concept map of two different languages be so similar as to fill all the same possibilities in the same way, with the same overtones and associations? There is no presumptive reason why an integral translation of even ordinary perceptions should be possible at all without ampliative commentary.

Returning to the question of what fractional numbers are, we can now see that it makes little sense to ask why the Egyptians never got around to doing things our way, in spite of the (to us) unquestionable advantages of our system. They said things they wanted to say in their own mathematical language, and if we wish to understand that language we are faced with the necessity of thinking our way out of our concepts of number and fraction and into those of the Egyptians, before we can compare the systems from a neutral standpoint.

Our concepts of number and fraction are now conditioned by the ongoing search for rigor and generality in modern mathematics. To make mathematics more exact, and to make it easier to get from one part of mathematics to another (from geometry to algebra, say), mathematicians since the seventeenth century have moved steadily away from intuitive or common sense definitions of the elements of mathematics, related to their role in calculations, toward definitions which allow the proper placement of the mathematical objects in question (frac-

Natural Knowledge in Preclassical Antiquity

tions, for instance) in a grand and unifying structure of axioms and postulates. The aim of reconstructing mathematics as a formal deductive system based on a few axioms and postulates is to allow great areas of mathematics to establish their logical consistency and general coherence with respect to one another.[14] Thus *A Guide to Modern Mathematics* (1964) defines a fraction as "a symbol in the form a/b in which a and b may be the names of any two numbers,"[15] where *Webster's Collegiate Dictionary* (1936) gave "one or more aliquot [dividing without remainder] parts of a unit or integer; the indicated quotient of one integer divided by another." In the newer definition all references to parts, units, integers, and operations of division have been expunged.

The modern fascination with the axiomatic method in mathematics which controls our view of numbers has historical roots in the geometry of Euclid and the subsequent development of Greek scientific mathematics. It is a tradition that Otto Neugebauer, the distinguished historian of ancient mathematics, has described as a "purely Greek development in a sharply defined direction" quite separate from the Babylonian [and Egyptian] traditions of practical measurement of quantities, areas, and volumes.[16] It made of Greek mathematics something close to philosophy: indeed, in the philosophy of Pythagoras and later of Plato, mathematics became the foundation of philosophy altogether, with the vision of a universe constructed out of the evolution of number.

The strong association between developments in logic and the axiomatization of many fields of mathematics in the late nineteenth and early twentieth century has given modern historians of mathematics a peculiarly Greek view of what mathematical advance consists of, and of the nature of mathematics as a creating, innovating, theoretical discipline, with close ties to physical sciences. These historians are perfectly well aware of this, of course. But it is essential to recall this historical heritage when trying to answer questions concerning the structure of a mathematical system and mathematical language constructed along different lines, and operating in a different cul-

Egyptian Fractions

tural context, as is the case with Egyptian mathematics, and in particular with something like unit fractions.

In Egyptian mathematics, Gillings wrote, "a fraction was denoted by placing the hieroglyph ○ ('r'), an open mouth, over any integer to indicate its reciprocal."[17] While unobjectionable in modern terms (which these are), this sentence imports into Egyptian mathematics the whole unifying and generalizing frame of modern mathematics, and leaves it as a foreign visitor in a civilization where it has no home. In Egyptian mathematics there is no concept of integer, or reciprocal, or even of a mathematical sign as a completely abstract operator.

The hieroglyph ○, which Gillings identifies as the sign for an open mouth, indeed has that function, but it is also the sound sign "r" or "er" for many words in ancient Egyptian, one of the most common of which is "er," a preposition meaning "into."[18] Thus the symbol ○ written over the number twelve in hieroglyphic characters signifies not "the reciprocal of the integer twelve" but "into twelve," that is, "into twelve parts."[19] Thus when Gillings writes, "When the Egyptian scribe needed to compute with fractions he was confronted with many difficulties arising from the restrictions of his notation," he announces only a partial truth.[20] The scribe is restricted not so much by his notation as by his concept of what a fraction is—and that concept is the representation of a certain measurable part of some thing.

The unit fractional notation is directly tied to the historical development of the mathematical system to serve as a tool of dividing and apportioning, gathering together and distributing, allocating and assigning parts of things, and *not* to the needs of mathematics as an abstract creative discipline. To the extent that the practice of mathematics was tied to this apportioning function, there was little conceptual flexibility in the system and no concept of number in abstraction from things numbered. An Egyptian fraction is a sign of something immediate and palpable—a twelfth part of a loaf of bread, a fifth part of a jug of beer—and has no independent existence in isolation from that material world.

Natural Knowledge in Preclassical Antiquity

The Origins of Unit Fractions

The concept of fraction in Egyptian mathematics as a part of a thing rather than the reciprocal of an integer was tied to its origins in measurement, from which it was never subsequently freed and abstracted. How this concept was first established remains in question, since many other measuring and counting traditions (the Babylonian, for instance) had no quarrel with the notion that if you cut off ¼ of a loaf, the remainder is ¾ of a loaf, rather than ½ + ¼ of a loaf. The question is similar to that concerning color words—why do different cultures divide up the spectrum in different ways and into different numbers of colors? Why do the English divide watercourses into rivers and streams based on their size, while the French divide them into *rivières* and *fleuves* depending on whether or not they flow directly into the ocean? The origins of such conceptual arrangements lie in the earliest strata of the experiences of a given language community.

That does not mean that their origins cannot be recovered; much of the history of science consists in striving to recreate the cultural circumstances under which such distinctions come to be made, in any period of history. Gillings suggests that the origin of the unit fraction indeed lies with problems of practical division. He raises the idea that in dividing commodities among ignorant workmen, say 3 loaves of bread 5 ways, under the unit fractional system each man would receive ⅓ + ⅕ + ¹⁄₁₅ of a loaf, rather than the results of the modern division, in which 3 men would receive ⅗ of a loaf, and 2 men would receive ⅖ of a loaf and ⅕ of a loaf, "which division might be regarded as an injustice by an ignorant workman."[21] I see no objection to this, though it depends on the assumption that Egyptian workmen were systematically more ignorant than modern workmen; in any case it is likely that the concept of partitions antedates the construction of the hieroglyphs that denote them.

In fact, it is probably here in the technology of partition and the establishment of equality of shares that the answer lies. I

Egyptian Fractions

suggest that the unit fractions are practical derivatives of sets of weights used to establish the value of objects of unknown weight in a pan balance. This suggestion goes some way toward giving a concrete answer to the problem, and at the same time explains the notation used and even the style of calculation typically employed in Egyptian mathematics.

Such pan balances, also known as assayers' or apothecaries' balances, are in use all over the world even today, along with sets of unit-fractional weights. One puts an object to be weighed in one pan of the balance, and adds weights of standard value to the other pan until the pans come to rest at the same height. The sum of the standard weights in the reference pan is the weight of the object in question. This happens to be the standard form of weighing things in ancient Egypt. Figure 2.1 shows the range of variation on the basic design, drawn from "Books of the Dead" from the second millennium B.C.[22] The balance had a religious significance as well: after death, one's heart was weighed in this very instrument against the feather of *ma'at* (virtue) and one's location in the Egyptian hereafter determined.

Many such balances and even more sets of weights to use in them, from the Egyptian Collections of University College London, were analyzed by Sir Flinders Petrie (the measurer of the Great Pyramid) in 1926. He discovered that among the weights of the Early Dynastic period (the time of construction of the Great Pyramid), only five fractional weights were usually found for each standard: $\frac{1}{2}$, $\frac{1}{3}$, $\frac{1}{4}$, $\frac{1}{5}$, and $\frac{1}{6}$.[23] These weights are of a variety of materials but fortunately some are of stone, and thus have not gained weight by oxidation. Using a balance from the predynastic period (before 3400 B.C.) with its set of red limestone weights, Petrie found it accurate to 1 part in 500.[24]

This extreme of sensitivity in a balance and its associated weights at a very early period seems to me a clue to the persistence of the bond between physical partition and the concept of fraction in the subsequent history of Egyptian mathematics. That the fractional weight set is predynastic indicates that the

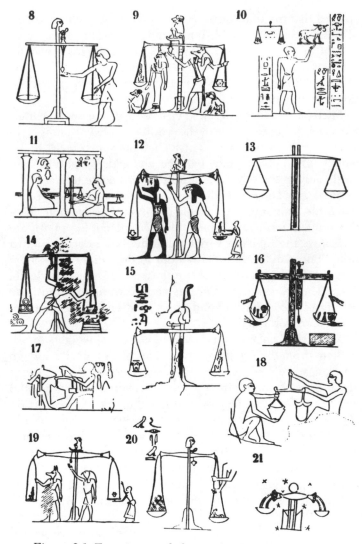

Figure 2.1 Egyptian pan balances from Ducros (1908).

Egyptian Fractions

physical foundation for fractional notation in weighing was
well established quite long before the tremendous growth of
organized distribution of goods and services to serve the state
machinery of the pyramid-building Old Kingdom. Unit-fractional weights that can exceed the limits of accuracy required
in practical affairs cannot but cement the idea of fraction as a
real part and forestall the development of fraction as an abstract
notion of partition in some general case. Where all the original
fractions are unit fractions (½, ⅓, ¼, ⅕, ⅙), the notion of fraction as real part must lead to a system in which all fractions are
unit fractions, no matter how large a number in the denominator, because the unity of the system of mathematics is under-
written at the deepest level by exact correspondence with acts
of physical measurement.

This correspondence might not be the case in every system
of social arrangements, but seems entirely plausible in a civilization in which, as in ancient Egypt, "craft numeracy" pre-
vailed. It sometimes escapes our attention, as members of a
civilization in which universal literacy and numeracy are practical requirements and legal obligations, that through most of
human history the ability to count beyond everyday household
need, and to write anything at all, were highly skilled and
laboriously learned crafts, on a par with specialities like car-
pentry, goldsmithing, and carving stone. Almost without exception, the documents from which we have deduced the character and extent of Egyptian mathematics are sequences of
solved problem-sets that owe their survival, via frequent copy-
ing for use in instructing new scribes, to the social realization
of mathematical calculation as a craft occupation.

Those trained in the craft of calculating in Egypt had to
memorize a very small number of rules for calculation with
halves, thirds, and sixths. Once these simple unit-fractional
equivalences were mastered, in conjunction with techniques
for addition, subtraction, multiplication, and division of whole
numbers and fractions, the scribes had recourse to compila-
tions of tables with which to solve and record problems of
greater complexity.[25] The enterprise is as different from our

Natural Knowledge in Preclassical Antiquity

characterization of the work of a mathematician as a public letter-writer of the nineteenth century is different from our concept of an author. This is not to say, or even suggest, that individuals of great talent did not arise and far exceed the everyday requirements of their craft, and in consequence produce mathematical constructions of great originality, power, and beauty, of which scant record survives. There are suggestions in the fragmentary correspondence of scribes, and in the character of some of the problems in surviving documents, of a genuine and familiar pleasure in the manipulation of number for its own sake. The difference between this quite predictable, if occasional occurrence of extraordinary excellence in the mathematical craft on the one hand, and the lack of a tradition of mathematical innovation, advance, and improvement on the other, is a matter of the historical sociology of science in this civilization.

Let us put the question, "Why was the unit-fractional concept never abandoned?," in another and more general way. Let us ask, "Under what circumstances do the styles of solution of problems in elementary mathematics change at all in any established mathematical tradition?" The answer, based on an examination of the whole history of mathematics, is that they change by slow and marginal drift and accumulation based on a repertoire of well-established concepts, *except* when mathematical innovation becomes socially institutionalized apart from the needs of practical calculation. This latter phenomenon has happened precisely twice: once in Classical and Hellenistic civilization, and again in its cultural stepchildren, Islamic mathematics and the European mathematics of the sixteenth century A.D. and after.

Most of the problems in the surviving corpus of Egyptian mathematics that can excite the aesthetic sensibilities of modern mathematicians are still problems within the context of practical calculation. These problems make up a class called "aha" or quantity calculations. They are like our linear (algebraic) equations with one unknown. An example is problem

Egyptian Fractions

26 from the Rhind Mathematical Papyrus. "A quantity and a fourth part of it give together 15." In modern notation, $x + \frac{1}{4} x = 15$. What is x? We solve this problem outside the boundaries of unit fractions. We say that $x = \frac{4}{4} x$, and $\frac{4}{4} + \frac{1}{4} x = \frac{5}{4} x$, and $\frac{5}{4} x = 15$. Multiplying both sides by 4, we arrive at $5x = 60$, and $x = 12$. The Egyptian solution proceeds as follows, in the text of van der Waerden.

> A quantity and a fourth part of it give together 15. "Calculate with 4, of this you must take the fourth part, namely 1; together 5." Then the division 15:5 = 3 is carried out, finally a multiplication, 4 × 3 = 12. The required "quantity" is therefore 12, the fourth part is 3, together 15. It is clear that the method followed here is that of the "false assumption." One starts with an arbitrarily chosen number as the required quantity, in our case 4, because this makes the computation of the fourth part easy. Four and a fourth part of four give 5. But the required result is 15; hence the quantity has to be multiplied by 15:5 = 3.[26]

What van der Waerden calls the method of false assumption, Gillings calls the method of false position, but the import is the same: the scribe falsely assumes the simplest number that could be chosen (in this case 4 because the fraction is $\frac{1}{4}$) in order to ease the labor of the solution.[27] What is of interest, from the present standpoint, is the way in which this technique of solution of "problems [that] could have had no practical applications" echoes the mechanical manipulations of the pan balance.[28]

The practical exigencies of the pan-balancing of an unknown weight require that one begin with a conjectural set of equivalents. It is as if a problem (object) looks as if it would be equivalent to $1 + \frac{1}{4} + \frac{1}{2} + \frac{1}{5}$ of the weight set. This estimation is the product of experience. But should it be in error, and the pan of balance weights fall below the common level required for equality, one pulls off the falsely positioned $\frac{1}{5}$, and substitutes the weight $\frac{1}{6}$. The clumsiness of the analogy

Natural Knowledge in Preclassical Antiquity

should not blind us to the identity of the concept of how solutions are achieved in both situations, and the extent to which the most interesting of Egyptian mathematical results, from the modern standpoint, bear the clear traces of their origin in the practical measurement of parts of things, both in substance and in procedure.

Technics and Civilization

The unit-fractional notation reflects the origin of the concept of fraction in the pan balance and associated weights. Survival of the notation and its validating concept were assured by the fundamental aims of mathematical work in Egyptian civilization—the collection and redistribution of goods in payment for services, in an economy in which money had no prominent place, but in which entitlements to payment in kind were differentiated to a fineness of degree that belies the notion that payment in kind is "primitive."

Consider the document in figure 2.2. It is a list of salary distributions from the Temple of Illahun, and dates from the Middle Kingdom Period (ca. 2400–1800 B.C.), the "classical" age of Egyptian civilization. There is no abstract reason why these unit-fractional salaries could not be recalibrated as common fractions, so that the temple worker who received $\frac{1}{4}$ + $\frac{1}{36}$ of a jug of Sd' beer (whatever that was) would get the same amount, distributed as 10/36 of a jug. But it is difficult to imagine why this should ever have been undertaken. The social inertia where such things as currency units are concerned is enormous, and resistance to change in common measures is even greater. Two hundred years after the invention of the metric system, in use everywhere else in the world, the United States clings resolutely to pounds and ounces, feet and inches, quarts and gallons. The Japanese, who could quite easily romanize their language (it is phonetically very poorly suited to Chinese characters), still require every Japanese child to learn to recognize 1,850 different written symbols, each with multiple pronunciations. And these things transpire in societies that otherwise have an almost uncanny thirst for novelty and

Egyptian Fractions

Personnel	Number of Portions 42	Loaves of Bread 1 $\bar{3}$	Jugs of Sḏ² Beer $\bar{3}$ $\bar{6}$	Jugs of Ḥpnw Beer 2 $\bar{3}$ $\overline{10}$	Corrected Ḥpnw Beer 2 $\bar{2}$ $\bar{4}$
The temple director	10	16 $\bar{3}$	8 $\bar{3}$	27 $\bar{3}$	27 $\bar{2}$
Head lay priest	3	5	2 $\bar{2}$	8 $\bar{5}$ $\overline{10}$	8 $\bar{4}$
Head reader	6	10	5	16 $\bar{2}$ $\overline{10}$	16 $\bar{2}$
Scribe of the temple	1 $\bar{3}$	2 $\bar{6}$ $\overline{18}$	1 $\bar{9}$	3 $\bar{3}$ $\overline{45}$	3 $\bar{3}$
Usual reader	4	6 $\bar{3}$	3 $\bar{3}$	11 $\overline{15}$	11
Wtw priest	2	3 $\bar{3}$	1 $\bar{3}$	5 $\bar{2}$ $\overline{30}$	5 $\bar{2}$
Imi ist c priest	2	3 $\bar{3}$	1 $\bar{3}$	5 $\bar{2}$ $\overline{30}$	5 $\bar{2}$
Ibh priests (3)	6	10	5	16 $\bar{2}$ $\overline{10}$	16 $\bar{2}$
Royal priests (2)	4	6 $\bar{3}$	3 $\bar{3}$	11 $\overline{15}$	11
Md ȝ w	1	1 $\bar{3}$	$\bar{3}$ $\bar{6}$	2 $\bar{3}$ $\overline{10}$	2 $\bar{2}$ $\bar{4}$
Thur guardians (4)	1 $\bar{3}$	2 $\bar{6}$ $\overline{18}$	1 $\bar{9}$	3 $\bar{3}$ $\overline{45}$	3 $\bar{3}$
Night watchmen (2)	$\bar{3}$	1 $\bar{9}$	$\bar{2}$ $\overline{18}$	1 $\bar{2}$ $\bar{3}$ $\overline{90}$	1 $\bar{3}$ $\bar{6}$
Temple worker	$\bar{3}$	$\bar{2}$ $\overline{18}$	$\bar{4}$ $\overline{36}$	$\bar{3}$ $\bar{4}$ $\overline{180}$	$\bar{3}$ $\bar{4}$
Another worker*	$\bar{3}$	$\bar{2}$ $\overline{18}$	$\bar{4}$ $\overline{36}$	$\bar{3}$ $\bar{4}$ $\overline{180}$	$\bar{3}$ 4
Totals (clerk)	42§	70§	35§	115 $\bar{2}$#	
Totals, without another worker	41 $\bar{3}$	69 $\bar{3}$ $\bar{9}$	34 $\bar{3}$ $\overline{18}$	115 $\bar{6}$ $\bar{9}$	114 $\bar{3}$ $\bar{4}$
Totals, including another worker	42	70	35	115 2	115 $\bar{2}$

* Omitted by the clerk. Or perhaps there should have been two temple workers.
§ Clerical errors, but only if there was in fact only one temple worker.
# If there were two workers, as seems most likely, then this should be 116 $\bar{5}$. Obviously the clerk did not add up all the fractions. He knew what they *ought* to total, and so he just wrote the numbers down without checking.

Figure 2.2 Salary distributions from the Temple of Illahun, Middle Kingdom (2400–1800 B.C.). Reprinted from Richard Gillings, *Mathematics in the Time of the Pharaohs* (Cambridge: MIT Press, 1972).

change. Egyptian civilization is celebrated, on the contrary, for its constancy and stability—in architecture, in sculpture, and in the fundamental technologies of settled existence. There was change, of course, but the pace of change was so slow in fundamental areas as to be invisible in a lifetime or even several lifetimes. Where the system of fractional numeration was built into structures of rights and duties that persisted for many

Natural Knowledge in Preclassical Antiquity

centuries, in a civilization with no larger appetite for novelty, the outcome was foreordained.

So Egyptian fractions survived for a number of reasons, not least of which was the absence of socially institutionalized innovation. Innovation and novelty are the stuff and substance of our technical civilization, the first civilization to institutionalize novelty. Only since the scientific revolution of the seventeenth century have decisive alteration and constant tinkering replaced slow drift and accumulated change. Novelty feeds on itself, since the faster things appear and disappear, the more difficult it is for them to become institutionalized, and the easier it is to change them.

Quite the opposite in Egypt. Since all higher learning was confined to a class of scribes who were at once the craftsmen of literacy and numeracy, the excellence of their craftsmanship actually forestalled mathematical invention. For in the performance of crafts, the skill resides in the user, not in the tool. Egyptian fractions look cumbersome to us, especially now that we have given over fractions to decimals on our calculators. But in the hands of a skilled user they were a powerful, adequate, and supple instrument. That the surviving Problem Texts are so cryptic, and require so much supposition to follow the path to the solution, is because the system employed human beings as prime movers in the intellectual sense, and here, as in the hauling of great blocks on sledges, the Egyptians made no deliberate effort to externalize the skills of the craft workers into machines.

Finally, there is the matter of aesthetics. While Egyptian fractions have been criticized for being cumbersome and inefficient and primitive, the real cognitive irritants that have driven historians to try to explain them, and explain them away, are that they violate our aesthetic criteria of elegance, simplicity, generality, efficiency, ease of learning, and ease of use. While these might pose as rational criteria whereby one might rank systems of mathematics, they are nothing of the sort. They are merely our preferences, and, in the end, to force them on a system whose beauty lay in the minds of its human

Egyptian Fractions

operators, is as wrong as it is futile. It is almost a kind of re-
venge on the Egyptians to thus criticize them for being the
proud and wise masters of their tools as often as we are the
abashed and foolish servants of our own, and for their failure
to play their assigned role in our dreams of cultural evolution.

▼ ▼ ▼

3

Hesiod's Volcanoes I.
Titans and Typhoeus

> Hesiod's *Theogony*, partly because it is the earliest surviving
> document of Greek literature devoted mainly to mythological
> topics, has occupied the most prominent position in many
> accounts of Greek myths, and so placed a strongly divine col-
> oring on the mythology as a whole. *G. S. Kirk*

Hesiod's *Theogony* has a special place in the study of Greek
myths, and the study of Greek myths has a special place in the
subject of comparative mythology. Students of other mythologies often make special efforts to find parallels in Greek myths
as a way of validating their conclusions, to the extent that they
share the conviction that Hellenic Greek thought is the well-
spring of Western civilization. It is canonically received that
the myths in Hesiod, in Homer, and in their survivals in Ionian
cosmology of the seventh century are the formative materials
out of which the Greeks made, for the first time in human
history, the transition from *mythic* to *rational* thought. As such,
Greek myths become more than another mythology: they are
the beginning of our culture, our arts, our sciences, and our
political forms. This explains why a document as otherwise
unexceptional as the *Theogony* is still studied and interpreted
at such length: as the first document of Greek mythology, it is
the beginning of the beginning of Western civilization.[1]

Hesiod's *Theogony* is a record of the lineage of the Greek
gods, and of their intergenerational struggles for power over
the cosmos. Most of the poem is as dynamic as the "begat"
portions of the Hebrew Scriptures, but when Hesiod chooses
to interrupt the steady thump of divine coition there are episodes of great narrative power. One of the most important of

46

Hesiod's Volcanoes I. Titans and Typhoeus

these is the battle between the Olympian gods and their allies, led by Zeus, and the preceding (Titan) generation of gods, led by Zeus's father Kronos. Another is the subsequent battle between Zeus and the monster Typhoeus. In these encounters the prowess of Zeus brings final hegemony of the Olympians over their earth-born predecessors, and Zeus's election by the other gods to be lord and king over them.

These episodes have been interpreted as representing, variously, an account of how order came to the cosmos, a mythologised account of the victory of the northern Aryan, sky-god worshipping Dorians over their southern non-Aryan, Titan-worshipping predecessors, and as an account of the origins of political society and ordered succession of kingship.[2] Indeed, any and all of these may be the case, and other plausible interpretations are possible.

The poem is very old, and contains stories known from Hittite and other records to date from about the middle of the second millennium B.C., so that some of the theogonic material (the succession of gods) is nearly a thousand years older than the earliest plausible date for the first written Greek text in the form we now have it.[3] The language of these battle episodes is also quite old, and contains poetic forms and usages and vocabulary found nowhere else in Hesiod or Homer.[4]

Regardless of how the battles between Zeus and the Titans and Zeus and Typhoeus are interpreted—as origin stories or aetiological myths—whether of cosmic order or local politics, and whatever they are to be taken to symbolize, their literal, superficial content is quite remarkable. Close attention to the sequences of events in the battles, to their appearance, sound, and their effect on the physical world, leaves no doubt that the phenomena described are volcanic eruptions. Not only that, but eruptions described so carefully and in such detail that the volcanoes in question can be identified and the particular eruptions of the volcanoes dated. The battle of Zeus and the Titans recounts the eruption of the volcano at Thera, in the Aegean Arc, commonly dated in the fifteenth century B.C. The battle of Zeus and the monster Typhoeus is an accurate representation of a violent eruption at Mount Etna in Sicily and is quite likely the great eruption of 735 B.C.

To extract such a conclusion from a poetical and mythological context involving the doings of gods and the gigantic figures who accompany them runs somewhat counter to prevailing conceptions of what myths are and what they express. While we increasingly explore mythologies of all cultures as sources of psychological data, and as material for the analysis of language and of thought process, it is quite definitely out of fashion to assert that they might have had as a primary function, at the time of their creation, the conveyance of explicit information concerning the natural world. While students of mythologies who employ the technique of structural analysis discuss the opposition of nature and culture or the wild and the civilized in myths, their interest is more with the presence of certain sequences of binary oppositions and their resolution than with any specific content: for such students the dialectical structure of myths is their most significant content.

While one cannot deny the power of formalistic analyses, this nascent orthodoxy is remarkably single-minded in its concern with the place a myth fills in a corpus of myths, even to the extent of reducing what the myth reports to a series of conceptual markers manipulated via systems of complex (if largely implicit) rules, the explication of which is the mythographer's true quest. Treating the generation of myths as a science, structural analysis definitely takes the position that what characterizes a science is not its content (changeable and evanescent) but its method. It may well be the case that the science of mythmaking and a modern physical science like chemistry are both best characterized by their rules of procedure. Yet chemistry is not about chemical rules, but about chemical substances and reactions; even so are myths not about their structures, but the matters which they discuss.

There is a generous portion of irony in all this. Structural analysis, as pioneered by Claude Lévi-Strauss in the latter part of the 1950s, was formulated in opposition to the then-prevailing notion that mythology is the product of the thought pro-

Hesiod's Volcanoes I. Titans and Typhoeus

cesses of the "primitive mind."[5] Rejecting the notion of a
primitive mentality supposed to be dominated by fears of the
unknown, structuralism in the hands of Lévi-Strauss's successors has in the end submitted preliterate thought to yet another
form of domination: possession by language. In this construction, myths become a series of experiments conducted by language in human hosts, for its own purposes.[6] It is in this dena-
tured form that the supposed origin of rational thought is retold
today.

Yet it is possible to recapture the natural-historical content
of the great battle episodes of the *Theogony,* with their strik-
ingly original blend of human, divine, and cosmic history. We
may set aside the literary and structural theories of myth to see
what this particular myth is about, and to argue implicitly that
the "strongly divine coloring" given to Greek mythology by
the *Theogony* is a consequence of how the *Theogony* was later
read, rather than of how it was written. This approach is more
accurate in its portrayal of what sorts of things concerned the
Greeks in the eighth and seventh centuries B.C., and it is also
more interesting.

Zeus and the Titans

Let us review Hesiod's account of the battle between Zeus
and the Titans as it appears in the best-known English translation of the *Theogony,* that of N. O. Brown (excerpts 630–719):

> For there had been a long war with much suffering on both
> sides and many bloody battles between the Titan generation
> of gods and the children of Cronus. The mighty Titans fought
> from the top of Mount Othrys, while the Olympian gods, from
> whom all blessings flow, the children of Cronus and fair-
> haired Rhea, fought from Mount Olympus . . .
>
> The Titans on the one side and on the other the children
> of Cronus together with the terrible monsters with their enor-
> mous strength, whom Zeus had brought from the lower dark-
> ness to the light. Each of them had a hundred arms growing
> from their shoulders and fifty heads on top of their shoulders

Natural Knowledge in Preclassical Antiquity

growing from their sturdy bodies. They grasped massive
rocks in their sturdy hands and took their place in the bitter
battle against the Titans.

On the other side the Titans prudently strengthened their
ranks. Both sides employed all the strength in their hands.
The limitless expanse of the sea echoed terribly; the earth
rumbled loudly, and the broad area of the sky shook and
groaned. Mount Olympus trembled from base to summit as
the immortal beings clashed, and a heavy quaking penetrated
into the gloomy depths of Tartarus—the sharp vibration of
innumerable feet running and missiles thrown. While the
weapons discharged at each other whistled through the air,
both sides shouted battle cries as they came together, till the
noise reached the starry sky.

Then Zeus decided to restrain his own power no longer.
A sudden surge of energy filled his spirit, and he exerted
all the strength he had. He advanced through the sky from
Olympus sending flash upon flash of continuous lightning.
The bolts of lightning and thunder flew thick and fast from
his powerful arm, forming a solid roll of sacred fire. Fertile
tracts of land all around crackled as they burned, and immense forests roared in the fire. The whole earth and the
ocean streams and the barren sea began to boil. An immense
flame shot up into the atmosphere, so that the hot air enve-
loped the Titans, while their eyes, powerful as they were,
were blinded by the brilliant flash of the lightning bolt. The
prodigious heat filled the Void. The sight there was to see
and the noise there was to hear made it seem as if Earth and
vast Sky above were colliding. If Earth were being smashed
and if Sky were smashing down upon her, the noise would
be as great as the noise that arose when the gods met in battle.
The winds added to the confusion, whirling dust around together with great Zeus' volleys of thunder and lightning-
bolts, and carrying the battle cries and shouts from one side
to the other, so that the uproar was deafening. It was a terrible
conflict which revealed the utmost power of the contestants.

Hesiod's Volcanoes I. Titans and Typhoeus

After many heavy engagements, in which both sides obstinately resisted each other, the battle was finally decided.

Throughout the bitter battle Cottus and Briareus and Gyges were in the forefront. They attacked relentlessly, throwing showers of three hundred stones one after another with all the force of their enormous hands, till they darkened the Titans with a cloud of missiles. Their brute force was stronger than all the valiant efforts of the Titans. They then conducted them under the highways of the earth as far below the ground as the ground is below the sky, and tied them with cruel chains.

Interpretation

There it is: a long war, a climactic battle, total victory for Zeus, total defeat for the Titans. There are classical and even Renaissance friezes and paintings of this episode with the Titans—giant beings with Michelangelesque torsos and snake tails—locked in violent combat with the Olympians.[7] The imagery of the battle is that of warfare before the advent of the hoplite phalanx of classical times. In this later period opposing armies formed in long ranks, their shields interlocked, and marched resolutely toward one another—the object of the battle being to break the coherence of the opposing rank, and then to pursue and slaughter a foe no longer protected by a shield-wall. We are no strangers to these battle tactics, evident in modified form at Waterloo, Gettysburg, and Gallipoli.

But this struggle of the Olympians and Titans is a melee—the opposing sides gather, hurl their throwing weapons, and then rush together shouting battle cries, and clash in a massive and deadly concussion. It is so easy for us to imagine the battle in this fashion, and to render it in terms of our experience of warfare, that we may forget to inquire what part of this picture is represented in the text—what was seen, and what was heard, and what was felt. Yet if we strip away later literary and artistic embroidery, something quite different can emerge. Let us consider the sequence as Hesiod reports it.

Natural Knowledge in Preclassical Antiquity

1. A long war has already been fought between the Olympians and the Titans before this engagement.
2. Both sides gather strength for a final encounter; Zeus's allies, the "hundred-handers," grasp massive rocks.
3. There are terrible echoes from over the sea.
4. The ground rumbles loudly.
5. The sky shakes and groans.
6. Mount Olympus trembles all over at the moment of contact of the opponents.
7. There are steady vibrations of the ground—like the stamping of innumerable feet running.
8. Weapons (i.e., massive rocks) whistle through the air.
9. Loud battle cries are shouted, reaching up to the high heavens ("the starry skies").
10. Advent of thunder and lightning signal the arrival of Zeus—a solid roll of sacred fire. Fertile fields crackle and burn. Forests roar with fire.
11. Earth and ocean streams and barren sea begin to boil.
12. An immense flame shoots up into the air, enveloping the Titans in a blast of hot air, apparent as a blindingly bright flash, and as prodigious heat.
13. The sight and sound are so enormous that one would think the sky had collapsed onto the earth and smashed it.
14. Arrival of windborne dust with lightning and thunder, with deafening uproar.
15. Titans are buried under a cloud of missiles and bound beneath the earth.

Hesiod has provided us with an interesting description, in which the principal elements are things felt—shaking and trembling,. intense heat—and things heard—terrible echoes, groans, thunderous cries, and a final deafening uproar. The visual aspect is limited to lightning, the immense flame shooting into the air, the boiling of the sea, and the arrival of windblown dust. The image of gigantic beings in contest is the poet's inference from what he can hear, feel, and see of the effects of the conflict on the world around him. He feels the

Hesiod's Volcanoes I. Titans and Typhoeus

ground shake, he hears a mighty roaring and the whistling of missiles overhead. He sees the onset of the lightning, and the culminating blinding flash, and then the confusion of dust. He cannot see the Olympians, the Titans, the hundred-handers, but he can hear them and feel them, and he can see some of the consequences of their struggle.

What results is neither naturalistic description nor personification, but a kind of middle ground in which the phenomena of the natural world are the results of the actions of *theoi*, the gods, but not the *theoi* themselves—and rendered comprehensible in this way. For this entire section is clearly something felt and heard and seen and then systematically interpreted and folded into the story of the succession of gods. The myth is the retelling of an observed and experienced sequence of events, interpreted as consequences of the doings of *theoi*.

Having made this simple claim—someone saw this, heard it, felt it—we can go no further until we can decide what, in our terms and from our naturalistic standpoint, this ensemble of phenomena could have been. We already know what the individual elements of the ensemble are in our terms because Hesiod's descriptions are already in our terms: explosions, fire, lightning, earth-trembling, blinding dust. The question is, rather, when we experience these phenomena in this order in the modern world, what do we call it, if we do not refer it to the actions and contestations of *theoi*?

What we call it in the modern world is a volcanic eruption. One need not know much geology to realize that nothing else happens on earth that could be described in this way. Knowing a little more geology allows one to make some inferences about the kind of eruption and the kind of volcano being described. Finally, some research into the roster of volcanoes in and around the Mediterranean known to have been active in the last few thousand years leaves but a single plausible candidate—the volcano at Thera (Santorini), which underwent a catastrophic eruption in the fifteenth century B.C.

One can anticipate at this point both Velikovskian cheers and Wilamowitz-Moellendorfian groans at the mention of

Natural Knowledge in Preclassical Antiquity

Thera. The well-documented, violent explosion of this island (located in the Aegean about 130 miles southeast of Athens and about 70 miles north of Crete, at latitude 36.404N and longitude 025.396E) in 1470 B.C. has become something of a *natura ex machina* for the ancient Mediterranean (see fig. 3.1).

It has been implicated in the destruction of Minoan Crete, identified both as the pillar of fire and smoke that guided the Jews out of Egypt and the producer of the tsunami that parted the Red Sea [sic], and has been suggested as the locus for Plato's Atlantis.[8] It is hardly original to suggest that it was an important phenomenon in the life of the ancient world. What I am suggesting is different, however. I am suggesting that a detailed record of the actual eruption survived to be written down by Hesiod. In other words, that the details of an event which took place in the middle of the second millennium were accurately preserved (whether in oral tradition in Greece or in written form as part of a Near Eastern theogony) for more than seven hundred years, and faithfully recorded as a part of the story of the origins of the world and the gods.

This suggestion raises obvious problems. The first is that such a transmission over three-quarters of a millennium implies a faithfulness in oral repetition or written record which many though not all students of myth would be reluctant to admit.[9] The second problem arises from the doubtful (to some) implication that one volcanic eruption can be distinguished from another on the basis of qualitative, colloquial, and figurative description, so that an identification can be made from the text nearly three thousand years later. The final problem, which enfolds the other two conjointly, is convincing one's hearers that this sort of transmitting of these sorts of things was one of the original functions of Hesiod's *Theogony* at the time it was written down, and establishes a direct linkage between cosmic and human history in that narrative.

The first and third problems cannot be settled here. There is too much at stake in tradition, reputation, erudition, habit, and ideology to make any headway on the questions of the nature of oral traditions, intercultural transmissions, and the

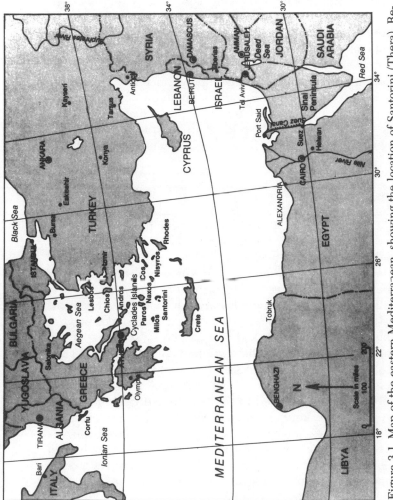

Figure 3.1 Map of the eastern Mediterranean, showing the location of Santorini (Thera). Reprinted from Fred M. Bullard, *Volcanoes of the Earth*, 2d rev. ed. © 1984, University of Texas Press. By permission of the author and the publisher.

function of myths written down. Yet the second problem is, in a sense, the key to the other two, insofar as the *Theogony* of Hesiod bears on the question. For there is no question that the battle of Zeus and the Titans is the figuration of an eruption, nor is there much question about when the eruption took place or what was erupting. Once the evidence for this is established, there is a prima facie case made for durable accurate transmission (problem 1) and the purpose of this kind of mythological recording (problem 3). It seems, then, that the best thing to do is to turn to the evidence linking the Titanomachy, as the battle is called, to Thera's destruction.

Mediterranean Volcanoes

Here we are at the crossroads of the sciences and the humanities, awash in distrust and mutual incomprehension. Yet it need not be so. First, the science we are to deal with here is an aspect of modern geology called, in a self-explanatory and straightforward way, volcanology. Geology certainly has its chemical, physical, and mathematical components, but it is still resolutely field natural history. That is to say, it is about particular things. Volcanoes are not like gas molecules—countably infinite in number and best treated by statistical aggregates and predictions. There are particular volcanoes. They have names and locations. As it turns out, each of the earth's volcanoes also has an "eruptive signature" all its own. This uniqueness is expressed in the pattern and frequency of its eruptions, the kind of material it throws out, and even the chemical composition of that material, which allows geologists to trace layers of volcanic ash on land or on the seafloor back to the volcanoes that spawned them, to compare ratios of specific minerals within other volcanic products and to use a variety of techniques to date the eruptions.

Since the end of the last ice age, about ten thousand years ago, there have been 5,564 identifiable eruptions by the 1,343 known volcanoes active in this period. Of these eruptions, 627 have specific dates assigned, with dates established by a number of means—historical, archaeological, geological, botanical, and physical (including radiometric dating via carbon 14 analyses).[10]

These eruptions may be graded in severity based on a number of criteria—the height of the volcanic cloud and whether it entered the stratosphere, the volume of material ejected, and other considerations, including descriptions of observers. These criteria are folded into something called the Volcanic Explosivity Index (VEI)—an attempt to do for volcanic eruptions what the Richter and Mercalli scales have done for the study of earthquakes. This explosivity index rates volcanic explosions with assigned numerical values, representing increasing violence, from 0 to 8. A score of 0 would be a nonexplosive gentle effusion of steam, gas, ash, or lava—a few thousand cubic meters—rising a few hundred meters over a volcano. A score of 8 would be a "colossal, paroxysmal, violent" event, ejecting not less than a million million cubic meters of material in a cloud that would shoot up more than ten miles and inject huge amounts of material into the stratosphere.[11]

In the last ten thousand years there have been no VEI 8 eruptions. There have been, of the 5,564 known, 3,913 of VEI 2 or less (mild to moderate). Of the remainder, 720 are VEI 3, 131 are VEI 4, 35 are VEI 5, 16 are VEI 6, and 1 is VEI 7. The scale, like the Mercalli scale, is logarithmic in character: each number on the scale represents an event ten times greater than represented by the previous number, and thus means an eruption ten times more explosive than an eruption rated one number lower. In turn, the higher the explosivity index, the less frequent the occurrence of such an eruption.

What about the Mediterranean? There are twenty-four volcanoes in Italy and Greece active in the last ten thousand years, eleven in Turkey, three in Iran, twenty-eight in Syria and Arabia. In terms of strength of eruption, most of these can be immediately eliminated. None of the Iranian, Turkish, or Syrian/Arabian volcanoes had an eruption with severity greater than VEI 2, with the exception of two VEI 3 eruptions in Arabia after 1253 A.D.[12] Of the remainder, the twenty-four in Greece and Italy, only Vesuvius, Vulcano, Stromboli, Etna,

Natural Knowledge in Preclassical Antiquity

Thera, and Kameno Vouno had recorded eruptions of VEI 3 or greater before 1500 A.D. The Vesuvius eruption of 79 A.D., which destroyed Pompeii and Herculaneum and killed Pliny the Elder (and quite a few others), was an eruption of VEI 5. Vulcano, in the Tyrrhenian Sea, had an eruption of VEI 4 in 183 B.C. Etna had three eruptions of VEI 3 or greater between 1500 B.C. and 735 B.C., and others of similar severity in B.C. 535 and 141. Kameno Vouno had an eruption of VEI 3 in 250 B.C. The explosion of Santorini (Thera) circa 1470 B.C. was the only VEI 6 in the last ten thousand years in the Mediterranean.[13]

So we are immediately down to two volcanoes, since only Thera and Etna had violent eruptions before the writing down of the *Theogony*. Vesuvius, Vulcano, and Kameno Vouno are strong enough, but too late in time to figure at all in the experience of Hesiod.

Etna is certainly a candidate, based on strength alone. It has had more than 200 identifiable eruptions since 10,000 B.C. (certainly a vast undercounting of the total), and 30 of these have been VEI 3 or greater. An explosion of VEI 3 ejects somewhere between ten million and one hundred million cubic meters of material, sends a cloud as high as seven miles, is severely and destructively explosive, and may shoot material into the stratosphere.[14] Nevertheless, there are two strong reasons for rejecting it as the candidate for the volcano described in the Titanomachy. The first is that the eruptive pattern described in the Titanomachy is very different from that for Etna. The second has to do with the difference in scale between the two. At VEI 6, Thera was somewhere between 1,000 and 10,000 times more powerful than the greatest recorded eruption for Etna.

We might best begin by giving a sense of the effects and appearance of such an event. The volcanic eruption and explosion of Santorini (Thera) is often compared with that at Krakatoa, an island between Java and Sumatra that was destroyed in a great caldera-forming explosion on the night of August 26 and the morning of August 27, 1883.[15] A *caldera* is a generic term for a large crater created by the explosion and subsequent collapse of a volcanic complex—Krakatoa, Thera, and Crater Lake in the United States are all calderas. A brief summary of the effects of Krakatoa will give us the proper scale with which to evaluate the effects and impressions created by the explosion of Thera.

Krakatoa had last erupted previously in 1680 A.D. In May and June of 1883, tourists visiting the island noticed steam issuing from the crater of Perboewatan, one of the three active volcanoes in the complex, and the presence of a fine ashfall that had killed most of the vegetation. On August 11, the last visitors to this uninhabited island noted that there were three vents active.[16]

At 1:00 P.M. on August 26, 1883, there was a series of tremendous explosions: a black cloud rose over the island to a height of 17 miles, and was followed by more sharp explosions and a tsunami. Explosions throughout the night were accompanied by severe air shocks. No one could sleep within a radius of 100 miles, and in Batavia, at that distance from the volcano, residents reported later that the windows rattled as if an artillery barrage were in progress.

Between 4:00 and 6:00 the next morning a series of tsunamis spread out, and at 10:00 A.M. the island detonated completely: the ash cloud rose to a height of 50 miles, and the sound was heard clearly four hours later on the Indian Ocean island of Rodriguez and in Central Australia, 2,900 and 2,200 miles away, respectively. "The intensity of the sound is better appreciated if one assumes that were Pike's Peak to erupt as Krakatoa did, the noise would be heard all over the United States."[17]

The immediate sequelae were disastrous. A tsunami reaching 120 feet swept Java and Sumatra, destroying 295 towns and killing 36,000 people; the wave eventually travelled all the way around the earth. A Dutch warship was carried one and one half miles inland. The ashfall created complete darkness for 22 hours in a radius of 130 miles, and at 50 miles, complete darkness for 57 hours. The shock waves travelling through the air from the climactic explosion cracked walls and windows more than a hundred miles away.[18] Dust fell days later 1,600 miles away. Where a mountain 1,400 feet high had stood, there was a hole 900 feet below sea level and two miles across. Explosions continued for two days and then ceased, and for three years the stratospheric dust created brilliant sunsets all over the world.

This eruption belongs to a category known as Plinian—named after Pliny, and by association after Vesuvius. Such eruptions begin with a series of premonitory earthquakes, felt locally and often over a long period of time, and then tectonic or deep-focus earthquakes felt over a wide area. Actual eruptions are explosive, and characterized by the ejection of ash, ignimbrite (fine, airborne rock material so hot that it tends to weld together as it falls and deposits), and other pyroclastic (fire-broken) products like pumice—a rock so filled with gas vesicles that it floats—rather than by lava flows.

When the eruption climaxes with the formation of a caldera, as in the case of Krakatoa and Thera, the roof of the magma chamber actually caves in. *Magma* is the name for the hot complex fluid from which volcanic gas, ash, pyroclastics, and lavas are all derived, and the magma chamber is the subterranean reservoir where this material is constrained under tremendous pressure. The collapse or rupture of such a chamber, if abrupt, can be catastrophically explosive. If the volcano in question is an island, and if the magma chamber is breached while the magma is still being erupted, the violence of the explosion can be many times magnified by the vaporization of inrushing sea-water coming into contact with the incandescent magma, in a so-called *phreatomagmatic* eruption, akin to the explosion of a gigantic steam boiler.[19]

The extensive geological investigation of the eruption at Thera has allowed a reconstruction of the course of the eruption of 1470 B.C. Eruptions at Thera (in recorded history) are always preceded by many months of strong, deep-focus earthquakes, and the strength of the earthquakes correlates closely with the violence of the eruptions.[20] A detailed analysis of the epicentral distribution of forty-four recorded major earth-

Hesiod's Volcanoes I. Titans and Typhoeus

quakes at Santorini/Thera (Mercalli 5 or greater—strong enough to wake most people from sleep) from 1500 to the present shows a high level of seismic activity for months and even years before a major eruption.[21]

The great eruption commenced with the ejection of huge quantities of pumice and ash; deposits from the first phase left up to 3 meters of pumice over the area.[22] The culminating explosion in the series began with the ejection of a huge, vertical "Plinian" cloud, perhaps 30 kilometers high. The ejection of huge amounts of pumice—more than 30 meters in thickness—created a tremendous overburden on the volcano, already punctured by a number of vents and weakened and fractured by the explosions. This led to a climactic, phreatomagmatic collapse and detonation in the course of an hour, followed by the ejection of 35 cubic kilometers of ash. This ash cloud left deposits on the Mediterranean floor more than 220 centimeters thick, and spread out over 200,000 square kilometers. Its violent ejection into the atmosphere caused catastrophic precipitation—the marks of flooding are evident as erosion of the ash deposits themselves—and accompanying darkness, wind, and dust storms. The total volume of collapsed material (as opposed to ejecta) is estimated at 60 cubic kilometers. While assessments of the violence of the explosion vary, it is generally believed to have been more violent than Krakatoa, and perhaps as much as three times as violent.[23] If so, it would make Santorini one of the half-dozen most explosive events of the last ten thousand years. We must remember, in assessing the impact on observers, that this was no uninhabited Pacific island, but a thriving Minoan colony on an island little more than 100 miles from Athens and 70 miles from Minoan Crete.

Let us place the description of Hesiod and the volcanologists reconstructions of the Thera explosion in parallel and examine them.

Hesiod	Thera
1. a long war	1. premonitory seismicity
2. both sides gather strength	2. increase activity

Natural Knowledge in Preclassical Antiquity

3. terrible echoes over sea	3. first phase explosions
4. ground rumbles loudly	4. Tectonic earthquakes
5. sky shakes and groans	5. air shock waves
6. Mt. Olympus trembles	6. great earthquakes
7. steady vibrations of ground	7. earthquakes
8. weapons whistle through air	8. pyroclastic ejecta
9. loud battle cries	9. explosive reports
10. Zeus arrives: lightning, thunder, fields, forests burn	10. volcanic lightning, heat of ignimbrites
11. Earth and sea boil	11. magma chamber breach
12. immense flame and heat	12. phreatomagmatic explosion
13. sound of earth/sky collapse	13. sound of above
14. dust, lightning, thunder, wind	14. final ash eruptions
15. Titans buried under missiles	15. collapsed debris

We should note that there is a complete one-to-one correspondence with no missing elements and that they are all in the correct order. In the above sequence, there is an important element that bears discussion: this is the arrival of Zeus at the climactic moment. Zeus's weapons are thunder and lightning, and they appear in the poem as the instrument of the Titans' downfall and in the volcanic sequence at the onset of the Plinian plume that inaugurates the final series of explosive events, characterized by the greatest heat and force.

Any volcanic eruption that forcibly ejects large quantities of ash and gas at high velocity into the atmosphere will, of necessity, create turbulence and static electricity manifest as lightning discharge: this has been observed all over the world. But only with the investigation of the volcanic island Surtsey, which appeared off Iceland in late 1963, did a more compelling phenomenon find scientific explanation. Observers recorded large numbers of lightning strokes discharged specifically from the air above the volcano into the throat of the volcanic

Hesiod's Volcanoes I. Titans and Typhoeus

cone—and systematic investigation of this phenomenon revealed that in a volcanic crater flooded with sea water, eruptions of steam and tephra (pyroclastics) greatly increase the
electric potential gradient and eject a huge amount of material
carrying a large positive charge, which is then repeatedly discharged back into the mouth of the crater from which it
emerged, giving the appearance of lightning bolts being
hurled into the mouth of the crater across a distance of several
kilometers.[24] In other words, at the climactic breach of the
integrity of the magma chamber at Thera, huge volleys of volcanic lightning immediately preceded the final phreatomagmatic collapse of the caldera with its associated heat and
noise—and the end of the Titans, giving rise to the interpretation (in Hesiod) that Zeus's intervention was decisive. The
volcano exhausts itself and disappears beneath the ocean, the
Titans are bound beneath the earth—end of the Titanomachy,
and arrival of the hegemony of Zeus. Or perhaps, not quite.
Consider the following passage, which immediately succeeds
the above in the *Theogony*.

Zeus and Typhoeus

Hesiod here continues his account of the battles of Zeus (*Theogony* 820–1022; trans. Brown):

> After Zeus had driven the Titans from the sky, monstrous
> Earth gave birth to her youngest child Typhoeus, after being
> united in love by golden Aphrodite with Tartarus. Typhoeus
> is a god of strength: there is force in his active hands and
> his feet never tire. A hundred snake heads grew from the
> shoulders of this terrible dragon, with black tongues flicker-
> ing and fire flashing from the eyes under the brows of those
> prodigious heads. And in each of those terrible heads there
> were voices beyond description: they uttered every kind
> of sound; sometimes they spoke the language of the gods;
> sometimes they made the bellowing noise of a proud and
> raging bull, or the noise of a lion relentless and strong, or
> strange noises like dogs; sometimes there was a hiss and the

Natural Knowledge in Preclassical Antiquity

high mountains re-echoed. The day of his birth would have seen the disaster of his becoming the ruler of men and gods, if their great father had not been quick to perceive the anger. He thundered hard and strong, so that Earth and broad Sky above, Sea and Ocean-streams, and the Tartarus region below the earth, all rumbled with the awful sound. Great Olympus quaked under the divine feet of its royal master as he rose up, and the earth groaned also. The heat from both sides, from the thunder and lightning of Zeus and from the fiery monster, penetrated the violet deep and made the whole earth and sky and sea boil. The clash of those immortal beings made the long waves rage round the shores, round and about, starting a convulsion that would not stop. Trembling seized Hades, king of the dead in the underworld, and the Titans who stood by Cronus and who live at the bottom of Tartarus. When Zeus' energy had risen to the peak, he took his weapons, thunder and lightning and the smoking thunderbolt, and jumped on his antagonist from Olympus and struck. He blasted all those prodigious heads of the terrible monster and dealt him a flogging until he was tamed. Typhoeus fell down crippled, and the monstrous earth groaned underneath. Flame streamed from the once powerful potentate, now struck by lightning, in the dim clefts of the rocky mountain where he fell. Large tracts of the monstrous earth were set on fire by the prodigious heat and melted like tin heated in moulded crucibles by skillful workmen, or like iron, the strongest metal, softened by the heat of fire in some mountain-cleft, even so did the earth melt in the flame of the fire then kindled. Zeus, in the bitterness of his anger, threw him into the abyss of Tartarus.

The notion that Zeus must engage still another titanic figure immediately following his stunning victory offends against literature. What would we think of the *Iliad* if Homer had employed the Trojan horse twice? Of a Norse Edda with two Ragnaroks? Of a *High Noon* where more outlaws get off the train as Gary Cooper is leaving town? As a gesture of respect

Hesiod's Volcanoes I. Titans and Typhoeus

for Hesiod, many commentators have hurled this entire section into the Tartaros of "interpolation," as it spoils whatever narrative tension the poem owns.

From the standpoint of myth-as-literature, there are plausible reasons for doing this, but the theoretical and philological evidence for them are secondary to the instinct that a drama may have but one climax. There is something Sisyphean and un-lord-of-the-universe in Zeus's having to unpack the thunderbolts twice in succession to vanquish the same sort of adversary. And if Gaia (Earth) and Tartaros are both still around, what can prevent them from rearing a continual succession of such noisy and destructive brats, requiring Zeus to smite and smite again?

The universe that threatens to totter and crash here is not that of Zeus, but of the *Theogony* as a narrative of how order replaced chaos. Taking the last of my hypothetical questions quite literally, "what prevents the emergence of more disorder in the universe?," the answer is: nothing whatever. If new Titans arise, then Zeus must again unsling his arms to silence them. What goes to sacrifice here is only an excessively neat and tidy view of cosmic history in which culture succeeds nature, and in which a binary opposition of symbols comes to closure and stasis in an atmosphere of uncontested hegemony and permanence of a beneficent *novus ordo saeclorum*.

It is in consideration of passages such as the Typhonomachy, as this episode is called, that theories of the text win out most often over what is written in the text itself. For if we set aside commentators and analysts for a moment, and ride with Hesiod alone line by line, we can enter a world in which the death of one volcano is later succeeded by the awakening of another, and in which the sequence is played out again—a fiery Titan, born of the earth and the underworld, again threatens stability and the restoration of cosmic order again requires the intervention of the most powerful of divine protectors.

It is with greater respect for the author as reporter, and less concern for the author as dramatist, that we can find resolution of the textual problem. For while the story of Typhoeus is

Natural Knowledge in Preclassical Antiquity

superficially like that of the battle with the Titans, its details mark it as quite distinct from the former contest. Behind the schematic mapping of rumble and smoke succeeded by peace and quiet is another painstakingly detailed and accurate rendition of the activities of a quite separate but very potent Titan—Mount Etna in Sicily. Here again volcanology and not philology renders the decisive verdict. The tale is that of another volcanic eruption, and this time of a different volcano.

In fact, Typhoeus was widely supposed by Hellenistic writers to lie below Mount Etna. The poem actually refers to "Aidnos" as the place where Typhoeus fell, and later manuscripts read "Aitnes." The language of the passage, while containing many un-Homeric and un-Hesiodic usages (i.e., not later formulae but unfamiliar), is such that "the close structural and linguistic similarity is only what we expect if both passages were composed by the same poet."[25] But the conclusive identification of Etna is in the faithful portrayal of the "eruptive signature" and physical description of that mountain.

Let us first, in a brief review, use elimination. There are six major volcanoes in the Mediterranean that have been active in historical time: Thera (Santorini), Stromboli, Etna, Vulcano, Kameno Vuono, and Vesuvius. Santorini was destroyed in the fifteenth century B.C. and did not reawaken until the eighteenth century of the present era. Stromboli has been continuously active and does not "come into being." Kameno Vuono's only large eruption was in 250 B.C. Vesuvius is largely quiescent and has left no geologic evidence of a catastrophic eruption after that of 25,000 B.C. until the eruption described by Pliny in 79 A.D. Vulcano's eruption pattern is not like that of the text, as no lava flows develop; instead, ash and great solidified bombs are ejected. So if we follow the text, the only candidate remaining is Etna. Let us examine and compare the text description with the known eruptive signature of Mount Etna.

A hundred snake heads grew from the shoulders of the terrible dragon, with black tongues flickering and fire flashing from the eyes under the brows of those prodigious heads.

Hesiod's Volcanoes I. Titans and Typhoeus

Etna is the largest of the Mediterranean volcanoes, rising to more than 3,000 meters above sea level. While the summit cone has remained active through the historical period, its flanks are dotted with hundreds of secondary and parasitic cones—more than 200 within twenty miles of the summit. These range in size from Mount Rossi, 450 feet high and two miles in circumference, to small "splatter cones" only a few meters across (see fig. 3.2).[26]

During a major eruption a large number of such cones may form or reawaken, sending out splatters of glowing lava visible as fire at night and as black against the daytime sky. The passage locates the dragon's head at the summit, and places the splatter cones correctly on the shoulders—glowing and flickering.

> And in each of those terrible heads there were voices beyond description: they uttered every kind of sound; they sometimes spoke in the language of the gods; sometimes they made the bellowing noise of a proud and raging bull, or the noise of a lion relentless and strong, or strange noises like dogs; sometimes there was a hiss and the high mountains re-echoed.

We think of volcanoes as producing lava but what we call lava is the extruded stony portion of the complex fluid we have already discussed as magma, which is also rich in superheated gases. It is gas pressure in the magma which builds up and drives the eruption; thus much of the work of a volcano is escaping gas, which makes a variety of noises, the analogies for which are dependent on the culture of the percipient. Paricutin (1943) in Mexico made a noise like a steam locomotive. Jorullo (1759), also in Mexico, sounded to the listeners like cannon fire. In the Mediterranean, Vulcano's 1786 eruption was described as accompanied by a great roaring; Stromboli (1952) gave out the sound of many rockets being discharged simultaneously.[27] The many voices of Typhoeus are the rumbles, hisses, bellows, roars, barks, and hisses of Mount Etna.

Natural Knowledge in Preclassical Antiquity

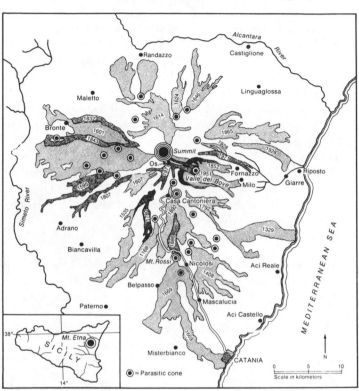

Figure 3.2 Mt. Etna in Sicily. Note the number of parasitic cones on the slopes. "A hundred snake heads grew from the shoulders of the terrible dragon, with black tongues flickering and fire flashing from the eyes" (Hesiod, *Theogony*, book 12). Reprinted from Fred M. Bullard, *Volcanoes of the Earth*, 2d rev. ed. © 1984, University of Texas Press. By permission of the author and the publisher.

Note that in Hesiod's description, only the hissing sound is described as echoing from the mountains: this is the signature of a major de-gassing episode, typical of Etna, and audible over great distances.[28] This accounts for the description of the dragon and the sounds of its activity, but we must also map the battle events onto their volcanic analogues, as in the case of Thera.

Hesiod's Volcanoes I. Titans and Typhoeus

[Zeus] thundered hard and strong, so that earth and broad sky above, Sea and Ocean Streams, and the Tartarus region below the earth, all rumbled with the awful sound. Great Olympus quaked under the divine feet of its royal master as he rose up, and the earth groaned also.

Subterranean roaring and groaning, rumbling and quaking; these mark the opening stages of an eruptive cycle. The quaking comes from shallow-focus earthquakes as the magma that will produce the eruption begins to rise up the central vent toward the crater from a reservoir several kilometers below the surface. Volcanic eruptions are associated with no less than three kinds of earthquakes: the largest are tectonic earthquakes with a deep focus—in the Mediterranean region this might be anywhere from 70 to 300 kilometers below the surface; these are felt over a large region. They often precede and follow and sometimes coincide with eruptions and are the result of major readjustments of the earth's crust. The other two types are volcanic earthquakes resulting from the ascent of magma (as above) or from the eruptive energy itself, as explosions during the eruption set off seismic waves in the crust. However different Typhoeus may be from the Titans, in this respect his encounter with Zeus must be the same.

The fire, the noise, the shaking continue as the battle is joined:

The heat from both sides, from the thunder and lightning of Zeus and from the fiery monster, penetrated the violet deep and made the whole earth and sky and sea boil. The clash of those immortal beings made the long waves rage around the shores, round and about, starting a convulsion that would not stop.

When the magma reaches the summit crater of Etna, the major phase of the eruption commences, showering lava "over the top"; jets as high as 200 meters have been observed. Enormous amounts of gas are released, leading to the same kind of volcanic lightning described at Thera. Eruptions of Etna can

Natural Knowledge in Preclassical Antiquity

be extremely violent; that of 1669 A.D. destroyed the summit cone at its climax. Heat, noise, the shaking of land and of the sea floor producing waves around the coast—all are visible around Etna, and in contradistinction to the climax typical of the caldera-forming volcano, such a phase might last for a long time ("starting a convulsion that would not stop").

> When Zeus had risen to the peak he took his weapons, thunder and lightning and the smoking thunderbolt, and jumped on his antagonist from Olympus and struck. He blasted all those prodigious heads of the monster and dealt him a flogging until he was tamed. Typhoeus fell down crippled, and the monstrous earth groaned underneath. Flame streamed from the once powerful potentate, now struck by lightning in the dim clefts of the rocky mountain where he fell. Large tracts of the monstrous earth were set on fire by the prodigious heat and melted like tin heated in moulded crucibles by skilful workmen, or like iron, the strongest metal, softened by the heat of the fire in some mountain cleft, even so did the earth melt in the flame of the fire thus kindled.

This passage confirms beyond a doubt that the volcano is Etna. An eruptive cycle at Etna begins at the top. The rising magma standing high in its conduit activates the summit cone and forms or reawakens the parasitic cones on its shoulders—the eyes and mouths of the monster. After a period of such activity which may last for months or even years, a climactic eruption takes place whereby a lateral eruption occurs as a fissure opens in the side of the mountain and lava pours out. As the summit conduit is rapidly drained, the lava pours out of an opening far down the mountain side. Thus one observes that the eyes and heads are all blasted (they go dark) just as flame streams from the fallen monster in a dim cleft. Now begins the lava flow and the setting of the earth on fire—here quite graphically described. The lateral fissure provides an outlet for gas as well as lava, and a major flow such as this typically brings an eruptive cycle to a close.

Where literary and other theoretical apparatus tell us that

Hesiod's Volcanoes I. Titans and Typhoeus

this is a doublet or an imitation of the Titanomachy, the text says nothing of the sort. It says that Zeus quelled another Titan in another manner in another place. Along the way the text provides a one-for-one correspondence between battle events and the eruptive signature of a known volcano, entirely distinct from any other in the Mediterranean world.

While it would be tempting to date this episode at 1470 B.C. (Etna had a major eruption at this time), this is historically not as probable as the eruption of 735 B.C., which was equally violent, and coincides with the arrival of Greek colonists in Sicily, and the establishment of colonies near Etna. Such dating can also explain the otherwise puzzling relocation of the capital of the colony from Naxos, founded by Chalcidians from Euboea under Theocles, to Leontini, far in the south of the Catanian plain, in the year 729 B.C. The colonists remained on the plain to obtain the benefits of the rich volcanic soil, but got as far away from Etna as was possible without leaving the region.[29] There is, however, the possibility that the eruption was that of 1500 B.C., which forced the relocation of the indigenous Sicels to the western end of Sicily. In any case, there is no doubt that it was Etna—one can only suppose that Hesiod would be more concerned about the establishment of Zeus's hegemony in the western Mediterranean than in the fortunes of antique *barbaroi*.[30]

Conclusion

I would like to offer a series of proximate conclusions and suggestions based on the above demonstration, which, whatever else it is, is certainly not a standard interpretation of these passages. Two of the most notable stories in Hesiod's *Theogony* are accurate accounts of natural phenomena, and a form of natural history. By carefully attending to the texts as we have them and reading what they say in the light of current geological, geographical, meteorological, and historical knowledge, we can identify the specific locations and dates of these events *and* by so doing we can help date the provenance of other portions of the work.

Natural Knowledge in Preclassical Antiquity

Reading Hesiod in this light should give us warning never to accept textual emendations of such passages made on purely philological grounds. We should learn to accept more freely what the sources report to us, since it is largely due to their uncritical repetitions that much of the material of interest in this regard has survived.

None of the ruling theories of mythology—literary, functionalist, structuralist, Freudian, Jungian, ritualist—is of much help with this aspect of mythic texts, and all such theories hold the study of myth hostage to opposed camps of specialists engaged in a power struggle over the origins and nature of human thought in a way which can obscure the literal sense of the texts. If we cart off the freight of excessive theory and take the texts at their words, we can develop a close connection in logic and content between parts of these earliest Greek texts and the speculations of the Ionian *physiologoi,* and perhaps come to a better understanding of the approach to natural phenomena that characterized Greek thought in the period down to about 500 B.C.

Anyone who wishes to study this material has an obligation to gain some knowledge of geology, geomorphology, volcanology, and a host of allied sciences—even if at an entirely popular and relatively superficial level—if he or she wishes to have access to the text's message.

Finally, no formalistic analysis of mythology, however polished, brilliant, time-tested, and attractive should ever be used to devalue the content of a myth until all possible avenues to a message of some sort being transmitted by that content have been exhausted. Even then the offending passage should never be struck out. We owe it to Hesiod to assume, in the first instance, that he knew whereof he spoke better than we. Neither can we forget that his world and construction of nature differ qualitatively from our own and had their own integrity and consistency.

4

Hesiod's Volcanoes II.
Natural History of Cyclopes

In the last chapter I explored the historical recording of volcanic eruptions in Hesiod's *Theogony*. In a world in which human and universal natural history interpenetrate, the story of these actual eruptions found its meaningful context (for Hesiod at least) in the story of the emergence of cosmic order. But this view of eruption-as-warfare does not exhaust the treatment of volcanoes in early Greek myth, and it allows us another look at the mythology as a form of natural history rather than imaginative or religious literature: the character of the beings called Cyclops (Greek *kyklopen*) also involves us in great natural events. The nature of these and allied beings has already been explored as a key to the nature of myth. In the late 1960s the distinguished historian of Greek thought, G. S. Kirk used his Sather classics lectureship at the University of California, Berkeley, to investigate the question of the nature of myths. Kirk's efforts resulted in the publication of *Myth: Its Meaning and Functions in Ancient and Other Cultures* (1970), a historical survey of approaches to myth, and a detailed exposition of the theories of the French mythographer and doyen of structural anthropology, Claude Lévi-Strauss. The latter's theories were, at that time, not yet well known in the United States, and had not been much applied to the study of classical mythology.[1] Indeed, the tendency to see the culture of the ancient Greeks not only as a subject for classicists, but also as a legitimate object for anthropological research, is a recent phenomenon in the Anglophone tradition. While anthropologists and sociologists have long devoted themselves to the study of the Greeks, British and American classicists have not been particularly interested in seeing the Greeks as a "people" in the eth-

Natural Knowledge in Preclassical Antiquity

nographic sense, in spite of the strong tradition of such work in France.[2] In pioneering this effort Kirk applied some of Lévi-Strauss's ideas (developed in a study of North and South American Indian mythologies) to Near Eastern and Classical Greek myths. Kirk then essayed some remarks on the relationship of mythic to philosophic thought, a subject to which he returned several years later in his *The Nature of Greek Myths* (1974).[3]

Within the ambit of classical mythology, Kirk chose to test the structural study of myths on an analysis of the figures of centaurs (half-man/half-horse) and cyclopes (one-eyed giants). He found Lévi-Strauss's analysis of myth, viewed as a problem-solving activity centered on the opposition of nature and culture, to be most illuminating with regard to these beings. Centaurs are indeed seen in classical accounts to exhibit both civilized and wild behavior; cyclopes provide an opposition of types and a variety of forms which constitute a linked series from wild, unpredictable, and dangerous, through tractable, and finally to "supercivilized." Lévi-Strauss's ideas about mythology were thus, for Kirk, in some limited sense confirmed by their application to these figures.

The theme of tension and opposition and the dialectical flavor of Lévi-Strauss's analysis was preserved in Kirk's treatment, as well as the idea that mythic thought is formal in essence and creates its own dynamic oppositions, while it is at the same time a conceptual technique aimed at fundamental questions. Kirk developed the opposition of nature and culture in the Greek world along the axis represented by the opposition of the Greek concepts of *physis* (nature) and *nomos* (convention), which captured not only the contrast wild/civilized, but the fixed conception of *nature* (as what is necessary or essential of a thing) over against the arbitrary character of civilized customs. In so doing, it is clear that Kirk found the older idea of nature myths, in which centaurs are representations of mountain torrents and cyclopes are figurations of other natural phenomena, to be an aberrant, nineteenth-century idiosyncrasy. For Kirk, all theories of unitary origins of mythology, whether Georges Dumézil's theory of an Indo-Iranian exem-

Hesiod's Volcanoes II. Natural History of Cyclopes

plar for every classical myth, or Victor Bérard's omnipresent Phoenecians, appeared as extreme as the Max Müller style of nature-myth interpretation.[4] Kirk was particularly critical of Wilhelm Roscher's lexicon of nature myths, the great turn-of-the-century thematic compendium that assigned myths to categories based on their representative function, and that made broad cross-cultural claims about the nature of mythologies.[5] Kirk dismissed such work as based on outmoded and condescending concepts of the thought processes of earlier civilizations, approximating the criticisms which Lévi-Strauss made of the concept of a "primitive mind" so fulsomely developed by Lucién Lévy-Bruhl.[6]

Kirk's essay on the cyclopes is, from the standpoint of classics, nevertheless refreshingly iconoclastic, and a principal subject was the redoubtable Samson Eitrem, who penned the article "Kyklopen" in the Pauly-Wissowa *Real-Enzyklopädie der klassischen Altertumswissenschaft,* the premier encyclopedic reference work of classical studies.[7] It was not the quality of Eitrem's scholarship that Kirk sought to question as much as the ideas behind his interpretations—and the tradition of nature-myth interpretation that he represented. Before moving on to what Kirk had to say, we must have a look at what Eitrem had provided on the subject of cyclopes, as a basis for discussion.

Eitrem tells us that the name *kyklopen* means, in Greek, "circle-eyes" or "round eyes." Hesiod, in the *Theogony,* says that they have a single eye in the middle of their brow, being like the gods in all other aspects, and that their strength, power, and skill are in their hands. Tradition specifies that the cyclopes aided Proitos in building the walls of Tiryns and Mycenae (which are still called the "cyclopean walls") and that they were the earliest inhabitants of Sicily, and co-workers of the forge of Hephaestos under Mount Etna; later Alexandrine and Roman tradition made them monsters inhabiting Sicily, Italy, and the Lipari Islands. There appear, wrote Eitrem, to be three unrelated groups of beings called *kyklopen* identified by the scholiasts: (1) the beings in Hesiod whose nicknames are

Natural Knowledge in Preclassical Antiquity

"Lightning" and "Thunder" and who give these weapons to
Zeus; (2) the builders of the walls of Tiryns, Mycenae, and the
other Mycenaean strongholds; and (3) the pugnacious, one-
eyed giant of the *Odyssey* and his fellow giants.

Looking at each of these groups in more detail seems to
make them even more disparate in form. The Hesiodic cyclopes are children of the union of Ouranos (heavens) and Gaia
(earth). They are the born before the "hundred-handers" and
after the Titans and are brothers to both these giant beings.
Freed from Tartaros (the underworld) by Zeus, they thank
Zeus by giving him thunder and lightning with which to battle
the Titans, and they themselves join the fray by hurling rocks
at the "hundred-handers." Apollodorus (I,1,4ff) says they gave
to Pluto his cloak of invisibility and to Poseidon his trident.

The cyclopes identified as the builders of the walls of the
Mycenaean strongholds are even more cloudy figures called
the *gastocheires, cheirogastores,* or *encheirogastores* (belly-
hands, hand-bellies) who are described as beings whose hands
emerge directly from their bellies. Nothing more is known of
them.[8]

Completely different from these two groups are the beings
of the *Odyssey.* These are mighty giants living far in the west
(the Mediterranean), in mountain caves on high peaks, living
alone, and a law unto themselves. The giant *kyklope* known to
us from the *Odyssey,* Polyphemus, was supposed to have lived
aloof even from other cyclopes, and was himself like a giant
mountain peak. This third group or category of *kyklopen* were
formerly neighbors of the Phaecians, but the latter were finally
forced to move away because of the cyclopes' unpredictable,
aggressive behavior.

Eitrem sees in the legends surrounding these groups an
association with fire-demons—this applies to the heavenly
lightning demons, the epic "trolls," and the wall-builders
alike. He sees in this the Indo-European tradition of fire-gen-
ius and metal-working genius as conjoint. He further stipulates
that the one-eyed giant is common in giant lore and that the
modern Greek folktale of the cyclops, whose eye spits fire, is

Hesiod's Volcanoes II. Natural History of Cyclopes

a version of the "Bösen Blick," the "evil-eye" conquered in this case by heroic Greeks who blind Polyphemus. Finally, he ropes in the wall-builders as *cheirogastores,* by arguing that their name is a misreading for *cheirogonos,* pointing to the phallic character of many fire-demons and technique-granting gods.[9]

So much for Eitrem's interpretations. Kirk, while depending on Eitrem for his descriptions of the provenance of the term cyclops, is not shy about his verdict. Eitrem engages, he says, in "exactly the kind of uncontrolled speculation I am anxious to avoid."[10] Eitrem did speculate in a manner typical of his time and training—in which analysis of myth was mixed up with a theory of folklore and folk and folkish things, which was at once both very literary (in a "symbolical" sense) and psychoanalytic in tone. For instance, the notion of phallicism carried with it the condescending judgment that the folk had their minds on fertility as a consequence of their closeness to the soil, and talked of sex and fecundity in all sorts of covert ways, and more often than a Wilhelminian sensibly found appropriate. This judgment carried with it more than a little *Hoch-Kulturlich* and professorial snobbery. One is drawn to agree with Kirk that we should avoid launching into such speculations. Nevertheless, we inherit a problem from Eitrem that has nothing to do with his interpretations: the apparent existence of three distinct types of cyclopes in the literary remains of Greek civilization. What are we to do about their differences and their connections if we do not accept the interpretation we have been offered? Here is how Kirk handled the problem.

At least this much can be said: The Cyclopes who were imagined as building the great walls of Mycenae were chosen, by a piece of popular lore rather than true myth, because they were giants. There is no special connection between these Cyclopes and the solitary giants visited by Odysseus, who were no extensive builders. Indeed, if the one type belongs primarily to folklore, the other to myth, no link beyond gianthood need be expected. Polyphemus professed no use

Natural Knowledge in Preclassical Antiquity

for the gods, but there are signs that he and his peers were favored by Zeus; this might connect them with Zeus's helpers, providers of thunder and lightning, who were specially rewarded in some way. As to these, "the gods themselves," we are largely in the dark; but for present purposes that does not much matter, since it is Polyphemus and his neighbors that chiefly concern us.[11]

This seems too draconian a solution. In effect we are told here that we can throw out one-third of the known kinds of cyclopes because they (the "wall-builders") violate a categorical distinction between myth and folklore, a distinction which is not by any means universally accepted, and is in any case several thousand years younger than the myths in question. We are then told that we can throw out another third of the cyclopes because, in spite of their connection to Zeus, we don't know about them, and because he (Kirk) doesn't want to talk about them and doesn't care about them. When he says "cyclopes" he means Odyssean cyclopes, and when he says Odyssean cyclopes, he means Polyphemus.

I think it is possible, however, to provide a coherent explanation of the various kinds of cyclopes which will allow us to understand the linkages between them, and which preserves all the fullness of their characters as given without indulging in speculation. While I share Kirk's skepticism for Eitrem's interpretation, I have the utmost respect for the power of the latter's erudition, and if he says that there are three distinguishable aspects of cyclopism, and three coordinate, widely developed traditions concerning them, then we ought to adopt an interpretation which covers all three forms if we are to understand the Greeks through analysis of their mythology. It seems self-defeating merely to discard the evidence that does not fit one or another useful scheme of interpretation—whether this be the distinction between myth and folklore, or the whole architecture of oppositions in structural anthropology.

When I read that a Mediterranean mythological tradition contains figures who are the sons of earth (Gaia) and sky (Our-

Hesiod's Volcanoes II. Natural History of Cyclopes

anos); who are like the gods but have a big eye in their fore-
heads; and who gave thunder and lightning to Zeus; figures
that are mighty, unpredictable giants who live in high peaks
in hollow caves from which they issue forth with sporadic violence that makes it dangerous to live next to them; figures that
live under Mount Etna and work the forges of Hephaestos, and
that also live in the Lipari Islands; that they are wall builders;
that they gave invisibility to Pluto and gave Poseidon the trident that shakes the sea; and that they are all associated with
fire demons, I think about volcanoes.

In developing the demonstration that each of these apparently disparate groups of cyclopes has a foundation in the nature of volcanoes, it is not necessary to attack or oppose my
predecessors. Neither Kirk nor I could have moved a foot without Eitrem, whose patient erudition is not at all out of date. On
the other hand, Kirk's analysis of Polyphemus, using structural
ideas, is very illuminating as far as it goes. But instead of studying these myths to determine the formalistic and formalizing
character of an opposition of nature and culture, or wildness
and taming, one might pursue those aspects of the myth that
appear to have a foundation in natural and human history, and
not in the abstract play of concepts. Embedded in the story of
the cyclopes is the historical experience of active, erupting
volcanoes (Etna, Stromboli, and Vulcano), quiescent volcanoes
which have erupted in the historical past (Vesuvius, the Phlegrean Fields), and a single volcano (Santorini-Thera) last active
in the middle second millennium B.C. The otherwise intractable problem of bringing together the three kinds of cyclopes
may best be pursued in this way—by seeing the term *kyklopen*
as representative of a diverse but related series of natural phenomena.

The most intractable group (of explanation) are the wall-
building cyclopes, the *cheirogastores* of Mycenae and Tiryns.
It would be handy if the "cyclopean walls" of Mycenae and
Tiryns were made of some volcanic rock, for then one would
have a connection between the power of volcanoes to build
walls of rubble, pumice, tephra, and other materials, and the

Natural Knowledge in Preclassical Antiquity

cyclopes who built the walls of legend. But Mylonas's archaeological account of the cyclopean walls of Mycenae indicates that they are made of limestone.[12] Limestone is *not* a volcanic rock. But when one examines photographs of the cyclopean walls of Mycenae and then travels to Mycenae and walks upon them, one sees that these large courses of stones are little more than a piled rubble of undressed blocks, dark in color, and with numerous pores.[13]

It would be most unwise to decide this question. However rather than reading modern mineralogical and petrological knowledge of which rocks are volcanic in origin and which are not back into the natural history of the first millennium B.C., one might suggest that it occurred to the Greeks to attribute these walls to cyclopes not only because they are so big only giants could do it, but because the dark, porous limestone of which they are made and the appearance of the rubble they create is superficially like volcanic rock. Not only are these rocks alike in appearance, but every volcano in the Mediterranean from Vesuvius to Santorini has limestone somewhere on it or near it: though not by our standards a volcanic rock, it is ubiquitously in the proximity of volcanoes. The similarity is enhanced when one sees the cyclopean walls in context on the hillside at Mycenae—sandwiched in between other levels like a series of superimposed but not concordant volcanic emissions typical of the region—rough, vesiculated stones jumbled together. In short, "cyclopean wall" might be not only a legend of giants in the earth who helped our forefathers but a naturalistic class name for volcanic rocks in a certain array. It is geology, but a different geology than our own—one that employs the character of external appearance, expressed in a different set of terms.

The origin of the name of this group seemed equally intractable—appearing in several versions—*cheirogastores, encheirogastores, gastocheires.* Hand-bellies? Belly-hands? It is difficult to follow Eitrem from *cheirogastores* to *cheirogonos* because they don't sound alike; the confusions which are most plausible are those arising from a scribal difficulty in rendering

Hesiod's Volcanoes II. Natural History of Cyclopes

some recollection of an oral tradition, or the perennial problem of faulty transcription. My guess, and it is only a guess, is that since the cyclopes are connected with the hundred-handers (*hekatoncheiron*), the element *cheir-* seems significant, and whoever collected the story and wrote it down wrote *cheirogastores* (hand-bellies) for *pyrogastores* (fire-bellies). *Pi* and *chi* are similarly formed and are easy to confuse if poorly written, but this is impossible to decide on if they are initial when sounded and we can make sense of the word either way. Even if one were stuck with *cheirogastores*, we might still make a plausible connection. If cyclopean walls are a description of volcanic rubble, or walls built from volcanic rubble, and we know that this rubble is ejected or thrown from the mouth of a volcano, then the person doing the throwing, the cyclops, must have hands inside its belly, and the connection between wall-building, hands, and bellies becomes somewhat clearer.

Let us turn our attention to the far-western Mediterranean, to the mountain-cave-dwelling, ornery, and unpredictable cyclopes we find in the *Odyssey*, and let us review their situation, following the presentation laid out by Kirk through the ninth book of the Odyssey (but with different intent). The cyclopes are

> an overweening and lawless folk, who, trusting in the immortal gods, plant nothing with their hands nor plough; but all these things spring up for them without sowing or ploughing, wheat and barley, and vines, which bear rich clusters of wine, and the rain of Zeus gives them increase. Neither assemblies for counsel have they, nor appointed laws, but they dwell on the peaks of lofty mountains in hollow caves, and each one is lawgiver to his children and wives, and they reck nothing one of another. [*Odyssey* 9:105–15]

One can see how ideal this section was for Kirk's purpose in explicating the structural wild/civilized dichotomy in laying out the absence of positive law, and affirming the cyclopes reliance on nature rather than culture, and their solitary existence. But there are other aspects of this narrative that reveal

Natural Knowledge in Preclassical Antiquity

another purpose. As book 9 unfolds, Odysseus and his men land on a fertile island and see the smoke rising from the land of the cyclopes. They row over to the cave of a monstrous man who lives apart from the others; he is not like a cereal-eating man but a solitary mountain peak. He too is lawless. Here Odysseus gets into his well-known troubles with Polyphemus; outwits him, blinds him, and escapes, though is nearly sunk when the blinded giant uses the sound of Odysseus's taunting voice to aim huge rocks, which he throws at the retreating ship.

The best way to show the connection between these characteristics and volcanism is with a map (see fig. 4.1).

This overhead view of the Bay of Naples contains all the described elements in their proper relations, and was, incidentally, chosen by Victor Bérard as the geographic setting in the Mediterranean most like that described in the *Odyssey*, in his now seldom-read *Les Phéniciens et l'Odyssée*.

The *campi phlegraei* or Phlegrean Fields (fiery fields) are a solfataric region—a region of volcanic activity characterized not by explosive eruption, but by emissions of gas. The setting is extremely parklike and fertile, the volcanic soil lending itself to easy cultivation of grain and vine. From across the bay, ten miles away on the island of Procida, Odysseus and his men see the "smoke" of the cyclopes rising on the mainland—the smoke and steam of the Phlegrean Fields. They do not venture there, but strike out toward the single giant's home—a giant like a mighty mountain peak, a lawless loner.

The single, dark solitary giant is the cone of Vesuvius. Polyphemus is not a "cereal-eating man": that is, no grain grows on his slopes, unlike the fertile Phlegrean Fields, which need not be plowed to be reaped. Here there are only wild goats and a dark cave. The contrast between the peaceable solfataric activity of the *campi phlegraei* and the brooding unpredictability and menace of Vesuvius is clearly stated. There is room here for the sequence of civilized, semicivilized, and wild which both Lévi-Strauss and Kirk find so useful, but it strikes me that this is less a conceptual scheme and metaphor—less

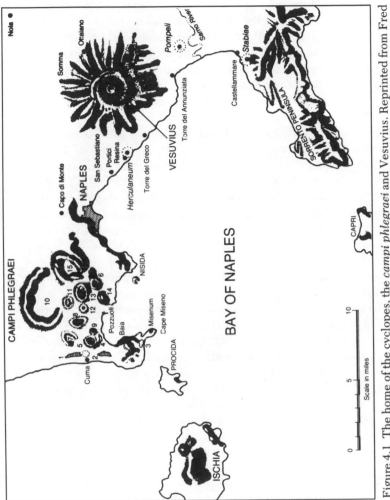

Figure 4.1 The home of the cyclopes, the *campi phlegraei* and Vesuvius. Reprinted from Fred M. Bullard, *Volcanoes of the Earth*, 2d rev. ed. © 1984, University of Texas Press. By permission of the author and the publisher.

83

Natural Knowledge in Preclassical Antiquity

a meditation on what it means to be civilized—than it is a nomenclature within the province of ancient natural history.

The third group of cyclopes bequeathed us, and the last with which we shall have to deal, are the figures of the *Theogony*, like the gods themselves but for the single eye. It struck me on reading this passage that it was an unwarranted assumption that the gods have two eyes and the cyclopes only one. What if, reading the passage as if all mountains were the abodes of the gods, the cyclopean mountains were those which rather than *no* eyes, had *one* single eye in the brow? Then, once again, the cyclops becomes an indicator of a volcanic phenomenon. As for the volcanic character of the Titans, hundred-handers, and cyclopes, we have determined this in the preceding chapter. The Titanomachy and the Typhonomachy are both volcanic eruptions and the mountains in question, Thera and Etna, have cyclopean associations. This is the least problematic of the associations. The other phenomena collected by Eitrem also fall into place. Cyclopes lived in the Lipari Islands (volcanic) and under Etna (a volcano). They gave darkness to Pluto and a trident to Poseidon—dark clouds and tsunamis or at least serious oceanic disturbances being common volcanic phenomena. A great cloud emerging from a volcanic cone might lead to the not terribly remote inference that a cyclops was a blend of earth and sky (Ouranos and Gaia). The poor Phaecians got tired of being erupted on, and moved away from their cyclopean neighbor.

As for their gift of thunder and lightning to Zeus, we have already discussed volcanic lightning, but there is a signal instance of the accuracy of this narrative material. The names of the three *kyklopen* who give thunder and lightning to Zeus are Brontes (thunder), Steropes (lightning), and Keraunos (strike or bolt). This has puzzled some interpreters, who wonder at the existence of three *kyklopen* to make two presents to Zeus—but we should recall that it is not self-evident that the flash of cloud-to-cloud lightning (so-called sheet lightning) is the same phenomenon as the bolt which is the visual component of cloud-to-ground lightning, and again we cannot expect

Hesiod's Volcanoes II. Natural History of Cyclopes

in every case a one-to-one mapping of their natural history onto ours. It remains the case, however, that for every aspect of the personality and physiognomy of the cyclopes, and for each of their legendary and mythic accomplishments there is a clear volcanic counterpart, of the correct and appropriate kind, in the right place.

In Plato's time, there was a debate over the nature of the old myths—the mythic heritage of Homer and Hesiod. While this focused mostly on the moral content of these myths, there was already the sense that these stories were irrational and false accounts of the world, which had to be replaced by rational and true accounts. Defenders of the old myths argued that they had a *hyponoia,* an underlying sense. Aristotle, more generous in the matter than Plato, suggested that there was an overlap in the early cosmogonies between rational and mythological explanation. Plato tended to reject the older myths as mere stories, and to employ myth in his dialogues in an entirely different way. In a subsequent essay (on Plato) I will take up this question in more detail, but something must be said here about the meaning and purpose of the *kyklopen* in the Homeric and Hesiodic myths, and in the legendary history of cyclopean activity which descends from preclassical times.

It seems to me that we have two polar alternatives to consider. One is quite traditional, and consists of ascribing to violent nature the agency of demonic forces—classes of beings whose depredations follow certain patterns. From this prospect one falls easily into the error of anthropomorphism—in which the giants, the titans, the gods, and the cyclopes are big men.

The other alternative is that all the activities of cyclopes and other classes of violent beings are personified in the purely metaphorical way, and this leads us to a view of myths as, in large part, literary embroidery and invention on a few themes of great interest. That is, counterposed to the view that the tellers and makers of these myths believed absolutely that the activities in question—storms, eruptions, and other attacks from nature—were the acts of hostile demons; we have a view

Natural Knowledge in Preclassical Antiquity

in which the whole thing is *ein Märchen,* a fairy tale, in the service of instruction, physical or moral. In either case, a myth is something which we must reject or at least demote beneath the level of the maturity of self-consciously rational discourse. If perhaps that antithesis seems too facile, we might agree that neither solution has much merit, though both views do at least accurately represent some ancient and modern views of what is going on in myths. Perhaps the problem can be phrased in a more constructive way?

What happens if one lives in a world that has volcanoes in it, in which there are eruptions of various kinds, producing various kinds of products of given appearance, and each having certain peculiarities and dangers associated? Some make gas, others throw rocks, others leak lava. Some are dangerous all the time, some intermittently, some never; you understand the connections and filiations, that these are connected phenomena. But there is no term for volcano, pumice, lava, eruption—that is, one does not yet have abstract conceptual vocabulary, no *science* on the one hand and *religion* on the other, with *history* somewhere else. One doesn't even live in a world where humans are "we" and nature is "it." Yet it seems important to record and discuss those observations in a way that exhibits the connections, because the connections are useful and even necessary pieces of information. This was the case in the highly volcanic Mediterranean Greek world of the seventh century B.C. and later. Nothing is lost if we make a one-for-one substitution of the Greek idea "cyclopean" everywhere we would say "volcanic," in order to explain how the three kinds of cyclopes are aspects of the same grouping of natural phenomena.

In other words, the cyclops material can best be understood as a taxonomy, an ordering of the world that employs a vocabulary for identifying and classifying types of living beings. This form of natural history served perfectly well to organize and categorize what seems to us to be disparate material but which in its own context has a principle of unity centering on volcanoes. In exploring the significance of such terminology at the

Hesiod's Volcanoes II. Natural History of Cyclopes

time of Homer and Hesiod, we need not see an evolutionary development from myth (as heuristic fiction) to philosophy, nor see it as any sort of a transitional vocabulary at all. It was the Ionian *physiologoi* who made a transition in the next century from natural history to natural philosophy. What was for Homer and Hesiod a natural sequence leading in cosmic but fully historical time from the origin of the world and of gods in unbroken sequence to the lives of men within the geographic compass of a known world, became for the Ionians a dynamic material process in a cosmological and timeless before.

It is when the function of mythology as natural history is forgotten or ignored that the material we have concerning the cyclopes seems demon-ridden, confused, and folkloric. Between about 800 and 600 B.C., the term *kyklopen* was adapted to conceptual use in an organizing vocabulary for volcanic phenomena in a historical setting. This conceptual residue remained in the *Theogony* and the *Odyssey* when a reified and less traditional vocabulary was developed, and in time both Hesiod and Homer began to seem part of the background from which they were among the first to emerge. This point could not be obvious, or it would be generally accepted, and one might inquire why the connection is not obvious.

In the first instance, I believe we are not yet shed of the predisposition to see the natural history of earlier epochs as primitive. While this assumption is almost inescapable and from a certain standpoint merited with regard to some of the physical sciences—dynamics for instance—it is not so for descriptive sciences. We have long since abandoned Aristotle's physics, while we continue to agree with him on how sharks give birth. There is a wisdom in assuming that the people whose works one studies are fully capable of describing their world until one knows enough about that world to assay what they said about it. In this case, consensus is forestalled by a disjunction in our intellectual world, not a mistake committed by the Greeks. The cyclopes have not been investigated for their connections with volcanoes because the specialists who know about the Greeks are trained in the humanities and not

Natural Knowledge in Preclassical Antiquity

the sciences. It is not just that they don't know about volcanoes, but because, as editors and translators of texts, they generally wish to render hexametric verse as poetic composition, and to see poetic composition implicitly in literary terms as an authorial and creative activity rather than a historical record—indeed, these texts and the use to which we generally put them invite such treatment.

Yet literary faithfulness to a poetic text may, in translation, manage to obliterate much of the material which is explicitly natural history and reportage. When Robert Fitzgerald's translation of the *Odyssey* says that Polyphemus did not eat "good wheaten bread" rather than that he was not a "cereal eating man" (i.e., a pastoralist rather than a cultivator), which is what the Greek says, we lose the intended contrast of soil fertility between the Phlegrean Fields and Vesuvius. We receive a sense that is metrical, epical, and pleasant to the ear, but not flexible in interpretation.[14] Lattimore has him as not an "eater of bread," Albert Cook has him standing on a mountain rather than being like one, Palmer has him "not like a man who lives by bread."[15] These descriptions are not wrong, but what is lost is the conceptual residue of the original and accurate Greek natural history. With the *kyklopen*, as in the case of Hesiod's volcanoes, a certain humility is definitely in order concerning the measure of our intellectual capacity to master the natural world, and that of those who went before—often long before us. It is so long since we came indoors and left the study of nature to specialists who speak to each other in languages only they understand, that we sometimes forget that the Greeks lived outdoors, in the natural world. Their mythology, at the time of its inception, had a substantial informative content concerning their environment and its history, which they shared and understood in a literal way. Not all myths are nature myths just as not all myths are origin stories, but many are, and many are both. While we debate the nature of myth we must remain forthrightly attuned to the real existence of myths of nature.

5

Thales and the Halys

Thales of Miletus (624?–545? B.C.) is traditionally the earliest of the pre-Socratic Greek philosophers, and therefore the first Greek philosopher.[1] This means that most histories of Western philosophy treat him as the first philosopher altogether. Both attributions are, strictly speaking, incorrect. Calling Thales a pre-Socratic philosopher is, in spite of our habits of use, like calling Thomas Aquinas a pre-Lutheran theologian. As for Thales being the first philosopher, the judgment merely repeats the Greek practice, after the manner of Aristotle, of ignoring all non-Greeks. Moreover, from the vantage of what we call Western philosophy, it is difficult to determine what was so Western about him, since he was a Greek-speaking Milesian with Phoenecian ancestors who lived his whole life in Asia Minor (in what is now southwestern Turkey).

So let us begin again. Thales of Miletus was one of the Ionian *physiologoi* (inquirers into "what is") of the sixth century B.C. In later tradition he was one of the "Seven Sages," or "Seven Wise Men," whom the Greek-speaking world honored as founding geniuses. This tradition credited Thales with a number of accomplishments, among which are the importation of geometry from Egypt, the prediction of a solar eclipse in 585 B.C., and the use of the constellation we call the Little Dipper (which has Polaris, the North Star, at one terminus) as a guide in open-sea navigation; the Greeks had previously restricted themselves to shore-hugging pilotage. He is supposed to have urged the coastal cities of Asia Minor (ancient Ionia) to federate and save themselves from conquest by mutual defense.

He is most often remembered for his philosophical cosmog-

Natural Knowledge in Preclassical Antiquity

ony: his theory of the origin of the world in the differentiation of the primal material, water. Since we do not possess any of Thales' writings, and do not know if he wrote anything we would recognize as a treatise, we know of these ideas and their fame as the first of this kind of thinking only through fragmentary references to him in the works of other and later writers. This is all well known and fully discussed in Kirk and Raven's *The Presocratic Philosophers*.[2]

Attached to Thales' name are some additional achievements. He is supposed to have fallen into a hole in the ground while walking along studying the night sky, thus becoming not just the first philosopher but the first "absent-minded professor."[3] Contrarily, he is supposed to have discerned by analysis of the weather that a bumper crop of olives was due, and to have cornered all the olive presses in the region, renting them out later at a handsome profit, and thus providing the first refutation of the formula, "Those that can't do, teach." Finally, Thales is credited with having diverted the Halys River in Anatolia so that the army of King Croesus of Lydia could cross over and invade Cappadocia. It would have been better for Croesus if he had had to turn around, but that is another story.

Thales' modern reputation rests on the intersection of two myths. The first is an ancient Greek myth about the origin of Greek accomplishment in thought, art, and political life through the agency of a restricted group of Wise Men. The second is the nineteenth- and twentieth-century European myth that the activity called philosophy supplanted the activity called myth in Greece in the sixth century B.C., and that this marked definitively the triumph of rational analysis and critical inquiry over storytelling and religious belief, once and for all in the history of the world. Further, this transformation is supposed to have happened only in Greece, and to have been autochthonous, with no connection with any contemporary developments in Phoenecia, Babylonia, Anatolia, Egypt, Persia, or any allied, contiguous, or subsidiary cultures.

Standard modern histories of Greek philosophy will typically disclaim this mythology, as, for instance, Guthrie's *The*

Thales and the Halys

History of Greek Philosophy: "We have outgrown the tendency . . . to write the history of Greek philosophy as if Thales
had suddenly dropped from the sky" (p. 39). Having dis-
claimed the myth, he then reasserts it, speaking of an "intellectual revolution" emanating from Miletus in the sixth century
in which "for religious faith there is substituted a faith . . . that
the visible world conceals a rational and intelligible order . . .
that autonomous human reason is our sole and sufficient instrument for this search" (p. 29). Miletus, he continues, lies at the
eastern fringe of Greek-speaking peoples and "had at its back
door the very different world of the East" (p. 31). This very
different world is soon characterized: "The Egyptian and Mes-
opotamian peoples, as far as we can discover, felt no interest
in knowledge for its own sake, but only insofar as it served a
practical purpose" (p. 34). This is in sharp contrast to Miletus
itself: "The environment of the Milesian philosophers . . . provided both the leisure and the stimulus for disinterested intellectual inquiry . . . Philosophy (including pure science) can
only be hampered by utilitarian motives" (pp. 30–31). These
and many similar passages (one could consult Bruno Snell,
Philip Wheelwright, Benjamin Farrington, Ernst Cassirer, H.
D. F. Kitto, J. P. Vernant, and many others for the same sort of
talk) confirm that the modern myth of Thales is still the origin-
story of secular rationality in Western civilization. It sets
Greek thinking over against all other and previous forms of
rationality, or of nonrational thought and discourse.[4]

The impulse to maintain this myth is very strong, for it has
survived the embarrassment of the demonstrated existence of
contemporaneous and robust traditions of rational inquiry in
the *Upanishads* of north-central India in the seventh century
B.C.[5] Outright deceit is averted with such formulae for Thales'
activity as "the birth of philosophy in Europe."[6] Without benefit of amplifying commentary or original italics, the qualified
phrase might locate (exclusively) the birth of philosophy in
Europe, or might merely announce its arrival in that geographical locale, without prejudice to philosophy's fortunes elsewhere. But the context is clear: what is meant is that Philoso-

Natural Knowledge in Preclassical Antiquity

phy was Born in Europe, and southwestern Anatolia is Europe as far as the eastern city gate of Miletus—beyond which lay a world mired in myth, fable, and practicality.

That there is no possibility that this myth (the origin of all philosophy in Greece) is true is almost beside the point: myths are stories that one embraces irrespective of their truth value—witness Tertullian's *certum est quia impossibile* (I believe it because it is impossible).[7] It is clear that the force of Thales' accomplishments, from the standpoint of the mythic history of philosophy in Europe is that he was the first exponent of disinterested inquiry into the nature of things, removed from practical concerns. The fame of Thales is decidedly in his thinking, and not in his doing. Moreover, it is in his thinking as a Greek, and not, as the Greeks saw him, as the conveyor of useful Egyptian thinking into the Greek world, which makes him the mythical progenitor of disinterested reasoning. Having set up thinkers as people who are not doers, historians of Greek philosophy, as mythologists of the transition from myth to philosophy, decline to see the originators of Greek philosophy as people interested in practical concerns, like calendrical reckoning and navigation, or commerce and military engineering. Having set the *physiologoi* up as the first Greek thinkers, they want the *physiologoi* to think, and they want them to think Greek thoughts in Greece. They are reluctant to believe that the earliest Greek philosophers went anywhere, and are particularly concerned to keep the *physiologoi* out of Egypt. Classicists, though geographically well-tutored and inclined to spend their university holidays going up and down the Mediterranean world, are remarkably resistant to letting certain well-known Greeks (Thales, Pythagoras) go anywhere during the historical horizon of the Greek Miracle, lest it be thought that they heard and brought home some non-Greek things.[8] Nevertheless, it is only about 400 miles from Miletus to the western delta of the Nile. In the summer the wind blows steadily from the north (the Meltemi or Etesian Wind). Making five knots—not an unreasonable speed before the wind—one could arrive in Egypt in three and a half days.

Thales and the Halys

The above reasons for these historical visa restrictions on Greek thinkers are not obscure. To the extent that early Greek thinkers were practical men with experience of metropolitan high cultures (Persia, Egypt, Aryan India), the mythic irruption of philosophy in Greece via a sudden sharp break or mutation of the modality of human thought gives way to the intellectual historian's stock in trade: the influence of one culture on another, and the especially interesting case of major influences converging from several sources at once.

But there is a deeper contradiction in the two myths of Thales reflecting the difference between what interested the Greeks and what interests moderns about the Greeks. The Greeks respected and expected plural excellence. That someone should be a good dramatist and a good soldier (Aeschylus), an original political thinker and a good geometer (Democritus), a lawgiver and a poet (Solon) was, if unusual in practice, nevertheless the cultural ideal. It was proper to accumulate *areté* (excellence) from several sources—in fact from as many as one might. The greatest heroes and the greatest sages, including Thales, are precisely these sorts of polymaths, and their legacy was more than the sum of their positive accomplishments: it was their exemplary significance in manifesting what humans might hope to attain.

In our culture, on the other hand, polymorph excellence is not only unusual, it is suspect. We are essentially and metaphysically committed to the division of labor, and on the choice of labor as expressive conjointly of psychological and spiritual make-up, innate capacities, and specialized training. Where the Greeks chose among possibilities, selecting several, we choose between alternatives, selecting one. For us the biologist who devotes hours every day to playing a piano in his laboratory is robbing biology, or music, or both. A physician who writes plays cannot, we suspect, be very happy that he chose medicine. We are permitted one serious excellence; everything else is a hobby. This does get in the way of our seeing Thales clearly, or seeing clearly any other culture that might have valued different choices.

Natural Knowledge in Preclassical Antiquity

Here, as is so often the case, Plato has had a good deal of influence upon the outcome of the debate. A disparager of polymathy and polymaths (he disliked Hippias of Elis), Plato also disdained getting one's hands dirty, at least as a habitual occupation. But even Plato maintains a sense of *sophia*—which we translate as wisdom, or as "skill" or "expertise"—as synonymous with *téchne*. Thus as late as the fourth century B.C. there was still a residuum of the early view. But following the Platonic line there are today elaborate theories of how improbable it is that someone (like Thales) could be good at a large number of things (making money, thinking, astronomy, mathematics, political theory, military engineering). Where there are not elaborate theories, there are extended arguments against the likelihood of this or that accomplishment being proven.

We want Thales as the first philosopher—and that means, by our cultural notions, that he cannot do anything else. When Bertrand Russell turned from the logical foundations of mathematics to a lifetime of crusading for educational reform, peace, and social justice, he was seen and he saw himself as having left philosophy behind. In the case of Thales, we dismantle his excellences: Hipparchus gets the astronomy, Euclid the geometry, Solon the politics, and making money and engineering are left unassigned. We feel close to the Greeks, and where we cannot be like them, we do what we can to make them like us. We explain away the Greek version of things. We say:

"The story is that Thales invented philosophy and developed geometry, and astronomy, and navigation, and made a shrewd profit in the olive oil business. These are all things the Greeks admired. But no single person could do all that. So Thales, while a historical individual, is also a mythic repository for the origin of a number of things that the Greeks liked. The story of Thales moving the Halys River for the army of Croesus is a case in point. It is exactly the kind of legendary accomplishment that ends up in the repertory of a hero, like Hercules, who diverted a river to clean out the stables of King Augeias as one of his 'seven labors.' Therefore," we say, "the story of

Thales and the Halys

Thales' moving of the river is probably false, as indeed are the vast multiplicity of his accomplishments." In other words, where the Greek myth of Thales comes into conflict with our myth of Thales, the Greek version has to give—and it has given.

There is a certain plausibility about this line of reasoning, especially if you see the Greeks at the dawn of civilization, not yet completely successful in parsing out myth from reality. But as we have seen in the essay on the cyclopes, this is a confused and hazy notion of myths and myth making, and of the contents of myths. But it is more than confused: it is paradoxical. We accuse the Greeks of making up a story about Thales to serve Greek cultural aims, and then we defend our accusation with a made-up story that serves our cultural aims, and is consonant with our metaphysics and, more frankly, our limitations. It is not that the historians and commentators of the modern period are not willing to allow the possibility that some of these accomplishments, at least, are real. The best, like G. S. Kirk, acknowledge Thales' activities "as a statesman and an engineer," noting that it is tempting to consider the Milesians "too exclusively as theoretical physicists."[9] Even when such acknowledgments are made, however, they are always acknowledgments of activities that show the versatility of men defined as thinkers.

In fact, Thales probably did do all the different things for which he was famous. That he did so only appears implausible if one insists on thinking of them as "all sorts of different things" that one person is unlikely to have mastered, and a person called a "philosopher" would be especially unlikely to master. The best way to make it clear that he probably did do all these things—and that doing them all was what made him great for the Greeks—is to take the most implausible, most clearly mythical, and work backwards toward the more plausible.

The outstanding improbability of his career is that, at the age of seventy-eight, he diverted the course of the River Halys in Anatolia, so that the army of Croesus could cross into Cappa-

Natural Knowledge in Preclassical Antiquity

docia. We get this story from Herodotus's *Histories*, Book 1. It is interesting that Herodotus, who believed almost everything that anyone ever told him, refused to accept the story. Here is what he says about the expedition of King Croesus.

> When he came to the Halys River, Croesus then, as I say, put his army across by the existing bridges; but according to the common account of the Greeks, Thales the Milesian trans- ferred the army for him. For it was said that Croesus was at a loss how his army should cross the river, since these bridges did not yet exist at this period; and that Thales, who was present in the army, made the river which flowed on the left hand flow on the right also. He did so in this way: beginning upstream of the army he dug a deep channel giving it a cres- cent shape, so that it should flow round the back of where the army was encamped, being diverted in this way from its old course by the channel, and passing the camp should flow into its old course once more. The result was that as soon as the river was divided it became fordable in both its parts. (Herod- otus *Histories*, 1:75)

Herodotus is wrong about the bridges; they were not built until the time of Darius, as a part of the Royal Road of the Persian Empire, which ran from Sardis in western Anatolia to Susa near the Persian Gulf—about 1,700 kilometers. But, in 547, when Croesus wanted to cross the Halys, he had to man- age without a bridge, and the nature of the Halys makes that a big problem.

It is a big problem because it is a big river. The Halys of antiquity is the Kizil Irmak of modern Turkey. It is about 1,600 kilometers (760 miles) long. It rises in the mountains of north- east Anatolia, and runs due west for some hundreds of kilom- eters, then turns northwest, passing not far from the modern city of Ankara on the east-central Anatolian plateau, and finally turns north, cutting through the coastal ranges and emptying into the Black Sea. Its full course is a large crescent, and it was the natural border between the ancient kingdoms of Phrygia and Lydia (to the west) and Cappadocia (to the east). It would

Thales and the Halys

seem the first order of business in testing a story about the fording of such a river is to look at the river, to find a likely place where it might have been forded. On the contrary, this seems to be an investigative strategy not yet tried.

The story of Herodotus indicates that the fording of the river by Croesus's army probably took place near the site where the bridges were eventually built, which puts it on the plain of Anatolia, near Ankara and the town of Kirrikale. An army marching east from Lydia is forced by geography into this crossing. To the immediate south are the Konya and the Tüz Golü depressions—poorly drained basins that are so water-logged that they cannot be farmed even today with much success. Moreover, were an army to march through this dismal swamp, it would have to surmount a thousand-foot scarp immediately to the east. If one marches south of the basins, one ends up on the Mediterranean coast and misses Cappadocia altogether, which is most annoying if one is trying to invade Cappadocia. On the other hand, one cannot go much farther north than the Ankara-Kirrikale passage without blundering into broken and difficult terrain.

Having got this close to specifying the crossing, it is difficult to be much more specific. For one thing, about 80 kilometers of the 150-kilometer stretch of the river where the ford might have been accomplished is today under water—the ancient bed drowned by hydroelectric power projects and their associated dams at Hirfanli, Kesikkoprü, Kapülükaya, and Bügara. Silting up of the river through two millennia of deforestation and topsoil erosion has also altered the landscape. Maps are not much help because the Turkish government is rather shy about releasing detailed topographical maps and air pilotage charts of this area, which is sensitive from a national security standpoint.

In the end, even a good map wouldn't be much help, because of the nature of the river. The river meanders. In fact, all rivers that have granular (as opposed to compact) beds meander, if they are not constrained by some topographic barrier. Meandering is the tendency of a river to travel in curves,

Natural Knowledge in Preclassical Antiquity

and follow a sinuous course, in preference to a straight line. Meandering is common, in fact, to many kinds of fluid streams. The Gulf Stream in the Atlantic Ocean and the Jet Stream, an upper atmosphere high-velocity current, both meander. Meandering streams typically shift their courses in a "meander belt" about fifteen to twenty times the width of the stream itself. The reason why a modern map would not be much help is that, as geological processes go, river meandering is very speedy indeed—and the course of a meandering stream can change dramatically in decades, let alone millennia.[10]

As a stream meanders, an interesting thing happens. The individual curves (the meanders) begin as slight bows, then gradually become more and more pronounced, and at a maximum look like ribbon candy. This is because the transported sediment and the pressure of the water "cut away" at the "neck" of each meander. Eventually, the neck is narrowed to nothing and the river cuts through it (see fig. 5.1).

This is repeated again and again as the meanders migrate downstream.

Once a river has cut through the neck of a meander, it flows for some time in two channels. Eventually the entrances to the longer curved channel silt in with fine particles deposited by the stream eddies, and the curved channel becomes what is known as an oxbow lake—named for its crescent form. Not all cut-off meanders are oxbows—there are longer and more sinuous cutoffs known as serpentines, or, in Australia, as *billabongs*. In areas that are well-drained and seasonally hot and dry (anyone who has ever been to Central Anatolia will attest that it fits the bill), it is not long before an oxbow lake evaporates and becomes an empty crescentic channel alongside a now-straightened river.

The story we have concerning Thales has two nested improbabilities, of which Herodotus was almost certainly aware, and which probably shifted his judgment in favor of bridges. The first is that traditional dates for Thales would put him in his late seventies at the time of this adventure. That he was able to do much significant digging was not likely. It would

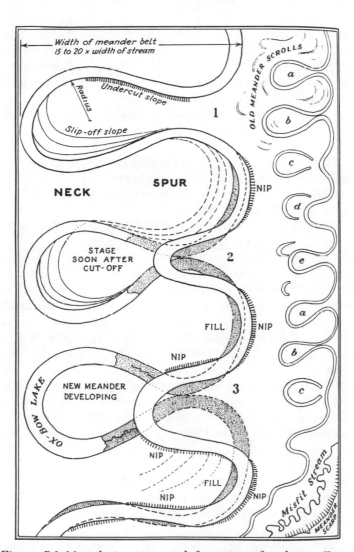

Figure 5.1 Meandering river and formation of ox-bows. From Armin Lobeck, *Geomorphology* (New York: McGraw-Hill, 1939).

Natural Knowledge in Preclassical Antiquity

have taken him a century, working alone, to dig a channel large enough to divert such a river. Clearly, if he did this thing, he directed the digging of a channel by others. The second is a question of time. The Halys is a very big river, and even with an army to do the work, digging a crescentic channel large enough to skirt an army encampment of many thousands would not have left enough time for a summer campaign. So Thales did the sensible thing. He found the dried-out channel of an oxbow lake. He moved the army within its radius. He had the army cut away at the silted-in entrances—first downstream and then upstream. The river divided its flow and became fordable in both its parts.

The Halys, though a big river, is the right sort of river to divert. One could divide the flow without having to dig a channel the full width of the existing river course and the diameter of the army encampment. The channel was already there, as were the natural cofferdams (the silt plugs at the ends of the oxbow), which needed only (though the engineering was not trivial) to be cut away. The question is not, by the way, whether the full channel could have been dug. It is a matter of time—whether it could be dug in time to complete the campaign.

We should remember that river management was the principal technological achievement of the most ancient civilizations—the Shang along the Yellow River, the Sumerian, Akkadian, and Babylonian along the Tigris and Euphrates, the Harappan along the Indus, and the Egyptians along the Nile. From the time of the earliest tool-making cultures (such as the Oldowanian of southern Africa, more than a million years ago), humans have settled in river valleys. The climate is milder, game and vegetation more plentiful, and transportation easier than in the mountains.

Some years ago, in a book with the now antique title *Oriental Despotism*, Karl Wittfogel argued that the edifice we call civilization—with its complex laws of persons and property—arose out of the necessity of large-scale coordinations of labor to keep from being overwhelmed by these rivers. In

Thales and the Halys

short, that the origin of civilization is hydraulic engineering. This includes the building of dikes, the digging of irrigation channels and ditches, the leveling of fields to receive diverted water, the mastery of plane surveying to re-plat seasonally for catastrophically inundated land, practical climatology and record keeping to predict the severity and timing of seasonal inundation, and a number of other associated skills (such as damming and gating, for instance), as well as the social technology of managing and coordinating the labor for these tasks. One need not accept Wittfogel's thesis (it is no longer generally in vogue) to acknowledge the historical foundation of his intuition of the hydraulic expertise of these civilizations, for it is amply documented in the historical record.[11] For the first three thousand years of recorded history, human beings were obsessed with rivers—depending on them, fighting them, praying to them, controlling them, worrying about what they were going to do next.

The Egyptians were particularly celebrated in antiquity for their management of and cooperation with the Nile. They could predict the date of its annual flood, plan for it, trap it, and resurvey all the arable land each year after it subsided. In the middle of the sixth century, in Thales' own lifetime, the Egyptian pharaoh Necho II, second pharaoh of the XXVI (Säitic) Dynasty, had undertaken to complete a canal begun by Rameses to connect the Nile with the Red Sea. (He failed, but Darius of Persia finished it about 520 B.C.)[12] In short, had Croesus's army needed to dig a channel, such a thing could have been accomplished with technologies contemporaneously well distributed in the eastern Mediterranean world. But Croesus didn't have to, because he had Thales.

One might ask, however, what Thales was doing with the army of Croesus anyway. Considering the errand Croesus was on—a major war designed to make him master of all Asia Minor—he hadn't much time for a philosopher. Not that kings didn't like philosophy. Frederick the Second of Prussia wrote pleasant music, poor metaphysics, and poorer poetry, and carried on a lively correspondence with the likes of Voltaire and

Maupertuis. But he did not invite them along for a chat on the
eve of the Battle of Rossbach. Croesus, too, would probably
have been more interested in sensation than reflection when
he retired to his tent. So Thales must have been there for
another reason.

It is a question, of course, of *whose* Thales he had along.
Our Thales, divorced from practicality and falling into holes
while dreaming about the cosmos, would have been less than
worthless to him. But the Thales who studied navigation, had
been to Egypt, knew geometry, and was known to mire himself
in practicality from time to time might have been a most useful
companion. Essential to this version of things is the datum
that Thales was from Miletus. Of all the Ionian city states in
southwest Asia Minor, Miletus was the only one not conquered
by the Lydians during the reign of Croesus's father, Alyattes.
Not that he didn't try—he assaulted them annually for eleven
years. When Alyattes died, Croesus made an alliance with the
Milesians as soon as he ascended the throne of Lydia.

While most anyone but an ancient historian hears of Miletus
only because Thales (and his successors) lived there, for 500
years before Thales' time Miletus had been a major trading
center. In the Greek expansion of the seventh century it sent
out numerous colonies. Milesian and nearby Carian mercenary
troops had been instrumental in helping Psamtik I of Egypt
(Psammeticus) regain independence from Assyria in the middle of the seventh century.[13] As a result of this circumstance,
the Milesians were allowed to found a trading post in the Nile
Delta. They were the first since the Bronze Age, and for a
long time the only Greeks so honored. By Thales' time the
settlement had become a thriving city—Naucratis—the principal link between Ionia and Egypt.[14] At this same time Milesian
sea power was the dominant force in the eastern Mediterranean, and it is no surprise that when Necho II, succeeding his
father Psamtik in 609 B.C., decided that Egypt ought to have
a navy, he had it built and manned by Milesians and other
Ionians.

It must be clear that this explanation leads, via this Milesian

Thales and the Halys

connection, to Thales' visit to Egypt. Thales was, according to his traditional life dates (and we have none better), about fifteen years old when Necho II ascended the throne of Egypt, started rebuilding the Red Sea canal, and ordered a navy from the Milesians. It was a glorious time in Egypt; it is known in cultural history as the Säitic Renaissance—when the Säitic pharoahs, Psamtik and Necho, made a deliberate attempt to restore Egyptian cultural and political prominence through a recovery of the learning and achievements of Egypt's classical past.[15] Much of what we know of the Old Kingdom of Egypt—the age of the Great Pyramids, of Khufu, of the Great Sphinx, of early Egyptian art, of mathematics and literature from the third millennium B.C.—we know via the historical researches, copies, and stylistic revivals of the Säitic Period. Egypt was an exciting place to be and the Milesians were there to share it.[16] There is neither a shred of evidence to indicate nor any rational reason to maintain that Thales never went to Egypt, and every reason to believe that he did. Given the volume of trade and commerce, one could get to Egypt over a weekend. It wasn't like scaling the Himalayas to Tibet or crossing the Gobi Desert or circumnavigating Africa (which happened in Necho II's time); it was simple. It is what Milesians did for a living. They were the official Greeks of the XXVI Dynasty.

That Thales had gone to Egypt to study was an essential part of his role as one of the Seven Sages, and was everywhere accepted in antiquity. No one in antiquity ever doubted it, because they knew that Egypt was, for Milesians at least, easily reached. When Thales was there he learned a good deal of geometry. He may have visited the works on the Red Sea Canal (it was under way, and everybody knew about it). He could have taken the first-millennial equivalent of the Grand Tour, visiting the high spots of the most distinguished high culture in existence. Nor does this rule out a trip to Persia, which was also undergoing a cultural renaissance, with scribes toiling away to learn the dead language, Sumerian, and restoring its classical literature with all the fervor that fifteenth-century Ital-

Natural Knowledge in Preclassical Antiquity

ian humanists displayed for translating Greek.[17] One merely
had to step out the back door of Miletus, as Guthrie has informed us, and one was in "Asia"—and Persia is in Asia by
this reckoning. There was much to learn, if one did not object
to being mired in practicality, and much to learn even if one
did.

But there is another reason why it is significant that Thales,
if he is to have accompanied Croesus to the banks of the Halys,
should have been from Miletus. Miletus is not just a port city
with access to Egypt; it is a river city, and the river that flows
by Miletus is the original Meander River of which Ovid wrote,
"The limpid Maeander sports in the Phrygian fields; it flows
backward and forward in its varying course, and, meeting itself, beholds its waters that are to follow, until it fatigues its
wandering current, now pointing to its source, and now to the
open sea."[18] Thales was from the city that has the river that
gives the phenomenon of "meandering" its very name.

He accompanied Croesus, then, not primarily for conversa-
tion with the chief after a hard day's march, but to superintend
the crossing of the Halys. He was a noted geometer (in its
original and literal sense of earth-measurer [geo + metron], a
surveyor, not a pure mathematician) who had been to Egypt;
he was a resident of a city allied to Lydia who knew about
waterworks; and he knew about meandering rivers. He was
there as a mercenary military engineer specifically to get the
army across the river, and perhaps to solve any other problems
that came the army's way.

This conjecture is a much better conjecture than the conjecture that he wasn't there and didn't do it, because what scanty
evidence we do have says he did do it—as indeed it specifies
that he went to Egypt. As well, it provides entree for seeing
his polymorphic excellence in its proper setting. Let us return
to my original contention that it is improbable that he was good
at all these different things only to the extent that they are
different things. They are not different things. They are all
modalities of mastery of water. If we can escape the notion that
philosophy is only philosophy when it is philosophy for its

Thales and the Halys

own sake, then we can see the unity of Thales' activities centering on water. He navigated across it, he moved it, he measured distances across it. He speculated on the causes of the Nile flood. He experienced water as a means of transportation, as a source of wealth (he knew to corner all the olive presses because it had *rained* a lot), and as an elemental substance, capable of a variety of forms, ubiquitous, but lacking any primary character; colorless, odorless, tasteless, essential for life. As a resident of a hydraulic civilization, nothing could be less strange than that he developed his skills and his ideas on the basis of water.

As for the birth of philosophy and the rational separation of man from nature, if such an event ever took place, it was a product not of abstract musing but of physical mastery of nature. From this standpoint, Thales' philosophical cosmogony was a speculative outfall of his practical work centering on water. The moving of the Halys, far from a spurious addition to a list of mythical accomplishments, was the final tour de force of his career—bringing together the experiences of a lifetime and a wealth of theoretical and practical skills—a clever and elegant solution to an engineering problem that only a Milesian geometer who had been to Egypt was likely to accomplish.

6

The True Identity of Soma

Indeed, if one accepts the point of view that the whole of
Indian mystical practice from the Upanisads through the
more mechanical methods of yoga is merely an attempt to
recapture the vision granted by the Soma plant, then the nature of that vision—and of the plant—underlies the whole of
the Indian religion. *Wendy Doniger O'Flaherty*

Somewhat more than three thousand years ago the Vedic religion of northwest India and the similar polytheistic religion of
neighboring Iran had as their central observance the drinking
of a beverage called Soma in India, Haoma in Iran. This beverage was prepared by pressing plants to produce an extract that
was then filtered and drunk by the celebrants. Descriptions of
the effect of this drink are a prominent part of the hymns of the
Rig Veda, which is the earliest surviving document of Indian
religion; they figure as well in the oldest stratum of the Zend
Avesta, the scriptural heart of traditional Iranian religion.
These texts report that the Soma/Haoma drinker was filled
with a sense of power, of joy, and of intense light, that fear of
pain and death vanished, that there was a sense of unity with
the gods and the universe.[1] Since this material dates from the
earliest stratum of both religions (Vedic and Avestan) we may
well ascribe it to that period of Indo-Iranian unity somewhere
around the middle of the second millennium B.C., as reflected
in the fourteenth-century B.C. Hittite inscription at Boghazkoi
in Anatolia, an inscription that extols Iranian (Mitra-Varuna)
and Vedic (Indra) deities as a composite group—the very deities implicated in the Soma/Haoma sacrifice.[2] These same texts
also indicate that sometime around 1000 B.C., the Soma/Haoma

The True Identity of Soma

plants from which the beverage was to be extracted became scarce. In the Brahmanas, the generation of Indian religious texts that follow the Vedas, we find listed a hierarchy of plants to be substituted if the real Soma cannot be obtained. In these Brahmanas, and in the portions of the Zend Avesta attributed to the prophet Zoroaster himself, there are also furious attacks on those who, in the face of the scarcity, would substitute in the Soma/Haoma ceremonies other forms of intoxication, particularly the drinking of alcoholic beverages.

The scarcity of Soma/Haoma was contemporary with a larger religious transformation (as a cause or consequence we do not know) that persisted in India and the Near East (including Iran) throughout the first half of the first millennium B.C.—that is, from 1000 B.C. to about 500 B.C. The transformation has a number of interesting aspects, the most striking of which is the rise of new religions: Upanishadic, Yogic, and Buddhist in India, Zoroastrian in Iran, and Judaic monotheism in Palestine. These religions have the following common features: abhorrence of drunkenness and religious intoxication, prescriptions of means to gain access to gods without such inebriation, recasting of the relationship of gods and men such that alienation, abandonment, or subjugation replace a former closeness, and finally alteration and even repudiation of existing priestly castes—whether Brahmanic or Magian. Moreover, it is from this period that we can date important myths—the story of Utnapishtim in the Epic of Gilgamesh, the Genesis story of expulsion from Eden—that center on loss of access to plants that confer immortality, or confer on the partakers godlike powers and perceptions.[3]

This same period saw a widening rift between Indian and Iranian religion. The pantheon divided in two, with the gods of one becoming the demons of the other. Further, in both traditions, the gods most closely associated with the Soma/Haoma sacrifice went into decline—Indra failed to make his way into the Hindu pantheon, and Mitra was rejected in favor of Ahura-Mazda in Zoroaster's recasting of traditional Iranian religion. The Soma sacrifice and the Haoma sacrifice remained

Natural Knowledge in Preclassical Antiquity

within Hinduism and Zoroastrianism, respectively, but it is clear that by the time of the Brahmanas, and certainly by the time of Zoroaster, the real presence (as it were) was a rare event, and the sacrifices had become elaborately ritualized. They were ossified and conventionalized memorials of a defunct live practice, bearing as little relation to the original as a Christian solemn high mass bears to the Last Supper. In fact, the ritual directions for the preparation of Soma in the Satapatha Brahmana are, in their fanatic pedantry, a scant cloak for the desperation of a priesthood whose stewardship of transubstantiation has failed categorically, and to whom Soma no longer shows his face. The incredibly elaborate directions for such non-Vedic elements as the ceremonial purchase of a cow to be exchanged in turn for Soma plants is but one of countless instances of a pharisaical response to the failure of rituals to work.

The Soma sacrifice is still performed today within the confines of Hinduism; among the remaining Zoroastrians there is a commemorative ceremony involving Haoma. In neither case is the original plant extract employed. Indeed, this was already clear by the late eighteenth century of our era, which marks the beginning of attempts by European students of comparative religion to find the identity of the original Soma/Haoma plant. Conversations with Brahmins conducting Soma sacrifices in the early nineteenth century indicated their awareness that Soma was no longer obtainable in its original form.[4]

Attempts to establish the botanical identity of Soma have continued to the present day without cease and without much success. The topic has been almost entirely the province of Vedic philologists, the guardian/interpreters of the classical heritage of Indo-Iranian antiquity. Like their counterparts the classicists, the guardian/interpreters of Graeco-Roman antiquity, they undertake many years of laborious and exacting study of the evolution of dead languages, and employ this knowledge in painstaking reconstructions of ancient religion, customs, and mores. Over the years the Vedic philologists' search for Soma has engaged the interest and aid of botanists.

The True Identity of Soma

But the problem of identifying Soma has remained open in spite of this collaboration. The problem of identifying the original Soma is succinctly stated by W. D. O'Flaherty, in her history of attempts to discover the identity of the Soma plant: "Few Vedic scholars knew any botany, and some of them may not have realized that they were dealing with a problem primarily botanical. The botanists on the other hand could not read the RigVeda, by far the most important source about *Soma*, and so they permitted themselves to enter upon speculations that often seem ludicrous in light of the Vedic hymns."[5] When scholars with an interest in the subject who were neither Vedists nor botanists, like Philip de Felice (in the 1930s) and R. Gordon Wasson (in the 1960s), took a position on the question, "professional scholars attached small importance to the theories of these 'outsiders.'"[6]

The "insider" theories of the identity of Soma have, since the middle of the nineteenth century, fallen into two groups. The first identified Soma with various species of *Sarcostemma* (= Asclepiads, related to the American milkweeds) or of *Ephedra* (known colloquially in the United States as "Mormon Tea," a mild stimulant) or as members of the genus *Periploca*, "all of them leafless climbers superficially resembling one another, yet belonging to genera botanically far apart."[7] This theory stems from the work of William Roxburgh who, in his *Flora Indica* (1832), identified *Sarcostemma brevistigmata* as *somaluta*, one of the Soma substitutes mentioned in the Brahmanas, and he further equated *somaluta* with the *som* plant of Charles Wilkins's 1784 translation of the Bhagavad Gita, which Wilkins took to be the original Soma. "For the next fifty years [to 1882] Sanskritists and botanists alike merely elaborated upon Roxburgh's identification."[8]

The second "insider" theory was that Soma was an alcoholic product of vegetable fermentation. It was revived in 1873 by Rajendra lala Mitra, who concluded that Soma was a kind of beer.[9] Interpreters of Soma as an alcoholic beverage also offered the opinion that it was a wine, perhaps made from rhubarb. The 1880s inaugurated a period of furious, inconclusive,

Natural Knowledge in Preclassical Antiquity

and redundant scholarly debate between these theories, most of the trouble resting with ambiguity of reference: did the argument refer to the original Vedic plant, to the substitutes in the Brahmanas, or to modern substitutes?[10] Both the alcohol theory and the *Sarcostemma* theory were fatally compromised from the start, although competent and even renowned scholars continued to recycle them for a century. The alcohol theory could not be true because the ritual directions for the preparation of Soma indicate that it is drunk immediately on pressing, and thus there is no time for alcoholic fermentation. The *Sarcostemma* theory is wrong, not merely because it is a common roadside plant whose milky shoots are sometimes used as a thirst quencher by travellers, but because it is explicitly identified as a Soma *substitute* in the Brahmanas and cannot be a substitute for itself. As early as 1855 Max Müller had inquired why, if *Sarcostemma* grows in Bombay, did the Indians of the Brahmanas use a substitute (the *putika* plant mentioned in the Tandya Brahmana, 9.5.1) for something so ubiquitous?

Before dismissing the insider candidates out of hand, we might speculate why, given their obvious insufficiency, they have remained popular for so long. One reason reflects the personal and theological background of "orientalists," who until quite recently were largely drawn from the ranks of civil servants, from students of comparative religion with strong religious commitments of their own (usually Christian or Hindu), or from the British, French, and German professoriate. All of the Hindus, and a large proportion of the Christian professors, were abstainers even from alcohol. The remainder, with the notable exception of the occasional outcaste opium eater, rarely tasted anything more potent than the college or club port, whiskey, or gin. For all of them, the notion of a powerful and mind-altering drug as the core experience of a religion implied (as in the orgiastic wine-guzzling Anatolian cults in Lydia and Cappadocia in 500 B.C.) a spiritual degeneracy which they, as respecters and even disciples of Indian religion, would be loath to attribute to the ancient Aryan devotees of the Soma cult. With no experience of drug-induced ecstasy, and with a

The True Identity of Soma

pronounced distaste for any homology between such ecstasy
and the fruits of legitimate mystical experience, they fell back
on interpretations of Soma that could satisfy the morphological
criteria of an ethnobotany that fit the remaining descriptions
of Soma substitutes, while not profaning the tradition by imputing degenerate excess.

It was an act of intellectual courage when in 1973 the Vedic
scholar Wendy Doniger O'Flaherty contributed her history of
the search for Soma to R. Gordon Wasson's *Soma: Divine
Mushroom of Immortality.* Without embracing Wasson's thesis that the original Soma was the fly agaric mushroom *Amanita
muscaria,* O'Flaherty made room for Wasson's speculations by
documenting with great brilliance the dead-end reached by
subscribing to either the *Sarcostemma* or alcohol theories, and
by giving in full detail not only their history but their etiology
as errors. Wasson, one of the outsiders ignored by specialists,
held an elaborately developed theory in which *A. muscaria*
figures as the prime suspect not only for Soma, but for the
drug of choice in many central Asian and Siberian shamanistic
cultures that employ such agents. Wasson's theory is the only
recent, comprehensive alternative to the traditional scholarship in this area, and he has developed his case with great care
and in full detail. His own experiences with hallucinogens—notably psilocybin and mescaline, undertaken as part
of his research into Amerindian drug-based religious experience—convinced him of the similarity of these experiences to
those described for Soma drinkers in the Vedas. These experiences, conducted in a scientific spirit, published in reputable
journals and in the company of distinguished psychopharma-
cologists, led him into the line of fire of scholars for whom such
speculations represented a profanation of the entire history of
mystical experience. When we recall that Aldous Huxley both
experimented with mescaline—experiments that lay behind
his *The Doors of Perception,* which extolled hallucinogens as
gateways to mystical ecstasy—and that he expropriated the
name Soma for the soporific, conformity-inducing drug of his
anti-utopian novel *Brave New World,* we can see the thickness

Natural Knowledge in Preclassical Antiquity

of the door slammed in Wasson's face. R. C. Zaehner, one of
the great modern students of Hinduism and Zoroastrianism,
and a fair-minded and sympathetic student of Indo-Iranian religion, undertook a refutation of Huxley in his *Mysticism: Sacred and Profane,* in which he denounced Huxley as a traducer
of one of the most valuable aspects of human spiritual experience. Zaehner, it should be noted, opted for rhubarb wine as
the Iranian Haoma.[11]

Wasson was mistaken in his identification of Soma as *A.
muscaria.* But Wasson was certainly moving in the right direction, for the descriptions of the experiences of Soma drinkers
preserved in the Vedas are worlds away from those produced
by tea or beer or wine: "it seems almost superfluous to point
out that none of the existing hypotheses is satisfactory. It is
difficult to imagine even very ancient and primitive peoples
becoming sufficiently excited about rhubarb wine to believe,
among other things, that it imparted immortality to the user."[12]

All attempts to determine the identity of the Soma/Haoma
plant, insider and outsider alike, have begun from the literary
evidence concerning the external character and appearance of
the plant; its size, shape, color, scent, and habitat, and whether
it had leaves or blossoms or seeds. The results have been
inconclusive. The descriptions of Soma are nowhere very clear
in the Rig Veda, and attempts to argue backward from the
Soma substitutes listed in the Brahmanas have led to identifications of Soma as wildly divergent as millet, afghan grape,
ephedra, horsetails, and rhubarb, no two of which are even in
the same botanical family.

There are several inferences we might draw. One is that
the Vedic, Brahmanic, and Avestan descriptions are not only
inadequate but now so confused with Soma substitutes and so
corrupted in transmission, using what is, after all, a highly
poetic diction and not a botanical nomenclature, that the textual descriptions are by themselves useless in locating the
plant. Or we might infer that the plant, already scarce in 1000
B.C., is now even more scarce and perhaps even extinct—if its
properties are as described in the Rig Veda one can well imag-

The True Identity of Soma

ine human beings hunting it to extinction. Again, we might humbly admit the possibility that it is out there and we have not located it: "Either the ancient hymns of the Rig Veda and the Avesta are gross exaggerations of fact or there grows in the vast mountain ranges of Northwest India a plant whose CNS [central nervous system]—stimulating properties, so well known to the old inhabitants, still remain hidden from modern man."[13] The fourth inference we might draw is that we have been going at the problem in the wrong way.

What if we should take the texts at their word and say that the external form and color of Soma vary, that the substitutes for Soma are even more varied, and that we may not even be talking about a specific plant at all, but about a pharmacological principle common to a number of plants? If this were the case, then instead of searching for Soma through the study of plant morphology we would take the search for Soma to be essentially a *biochemical* rather than a *botanical* problem. Wasson wrote that as far as he knew Soma was the only plant that humans had ever deified, and was therefore an anthropological as well as a botanical curiosity.[14] But if Soma were not the plant but the activity of the plant, then what was worshipped was not the plant itself but the divine activity of the god Soma in it.

When we look at the texts we find that while descriptions of the forms of Soma and its substitutes vary, the descriptions of its effects and the descriptions of the directions for its preparation are quite uniform. It is here that we should begin our search. Soma plant(s) must bring about certain effects; to bring about the desired effects (and to prevent undesired effects), the mode of preparation must be specific within a narrow range. We should inquire in the following order then: What were the effects? What sort of chemicals are known to produce these effects in humans? What sorts of plants contain these chemicals? Of the plants that contain these chemicals, which were available in India and Iran in the second millennium B.C.? Of the plants containing these chemicals that grew in India and Iran, which could produce a potent extract if prepared according to the instructions? If a variety of plants have the active principle, then the means of preparing them must extract it; additionally, if there are potentially dangerous side effects in the candidate plants, the preparation must eliminate them with a high degree of confidence.

The Effects

Several descriptions of the effects of Soma are to be found in the hymns comprising books 9 and 10 of the Rig Veda. Soma is said to impart a feeling of delight (9.56), of strength (9.67 and passim); it grants sage speech (9.67), great mental power (9.100), and an experience of pure light beyond life and death where all longing is satisfied. In 10.119, the "Song of the Soma Drinker," the celebrant feels himself rise above the world and worldly powers and is shot into the heavens, where he becomes as one of the immortal gods. In the Zend Avesta there is textual evidence that Haoma is not like other intoxicants: "All other intoxicants are accompanied by the fury of the bloody spear, but the intoxication produced by Haoma is accompanied by truth and joy; the intoxication of Haoma makes one nimble."[15] Haoma is the "god before whom death flees," an epithet common for Soma in the Rig Veda.[16] The Satapatha Brahmana declares: "Soma is truth, prosperity, light; and sura [a beer] is untruth, misery, darkness."[17] Objectively, none of these descriptions is diagnostic or of much help in isolating the specific drug. Such subjective responses might well be characteristic of any of the known potent intoxicants, including alcohol. The descriptions fit accounts of the effects of marijuana and its relatives; of opium, of peyote, psilocybin, mescaline, a variety of alkaloids, including reserpine, caffeine, nicotine, cocaine, or lysergic acid amides, and finally ibotenic acid (from *A. muscaria*). But even though the pharmacological activity of these drugs overlaps considerably, a number of them can be removed immediately from consideration. Peyote, mescaline, psilocybin, and cocaine are found only in the Americas, as is tobacco (nicotine). Caffeine and reserpine are all too mild as stimulants to receive much consideration. As much as one

The True Identity of Soma

might like tea, coffee, and tobacco, they do not offer intimations of immortality. Of the drugs catalogued above, only marijuana and its relatives, opium, ibotenic acid, and the plants that contain lysergic acid amides were botanically present in the India of the second millennium B.C.

Let us consider the active ingredients of each. Marijuana is a generic name for plants of the genus *Cannabis,* one species of which, *Cannabis indica,* was known in India in the second millennium B.C. It is a short, densely branched, leafy plant, rarely more than eight feet in height. Both the leaves and the resin from the flower buds are psychoactive, with the latter (known as *hashish*) considerably more potent. The leaves *(bhang),* the flower tops *(ganja),* and the resin *(charas)* may all be smoked or ingested to produce the desired effects. The active ingredients, called cannabinols, and especially THC (tetrahydrocannabinol), are strong soporifics that induce euphoria, drowsiness, increased physical appetite, transient memory loss, and a general feeling of well-being. The cannabinols are not alkaloids, and are not water soluble.

Opium is the name for the residue of the juice of the fruit capsules of the poppy *(Papaver somniferum),* which is harvested in liquid (latex) form from incised capsules and worked into a reddish-brown and bitter cake, which may be smoked or ingested, absorbed through the mucus lining of the mouth, drunk in a simple alcoholic tincture (laudanum), and in its purified active forms, morphine and heroin, may also be injected. As its species name suggests, it is extremely soporific, and the principal visionary effects (when present) generally occur in a dream state; it induces passivity and quiet, and is, unlike marijuana, strongly addictive in all its forms.

Ibotenic acid is the principal active agent of the plant *A. muscaria,* the fly agaric mushroom, a plant common throughout boreal, sub-boreal, and cool temperate Europe, Asia, and North America. Ibotenic acid is a water-soluble alkaloid that can be extracted from the crushed fruit body of the mushroom—especially the stem—and that produces hallucinations and delirium, but also produces headache, depression, nausea, confusion, convulsions, and other equally unpleasant side effects. Contrarily, it is reported to produce in the post-stupor phase of intoxication, a feeling of lightness, elation, physical strength, and heightened sense perception. "The effects are variable but perceived as stronger and more dangerous than those of psilocybin: at high doses delirium, coma, and amnesia have been reported."[18] It should be noted that *A. muscaria* contains only traces of the deadly poison muscarine, found in other Amanita species, notably *Amanita phalloides*. The hallucinogenic alkaloid muscimole, about ten times as potent as ibotenic acid, is probably not a constituent of live *A. muscaria* but is produced from ibotenic acid after the plant is harvested. This would explain the preference for ingesting dried specimens.

Lysergic acid amides occur in India as elsewhere as natural products of the life cycle of a fungus parasitic on grains and grasses. The fungus, known colloquially as ergot, consists of a number of species of the genus *Claviceps*, infecting a wide variety of domestic and wild species of grasses and cereals. Lysergic acid amides also occur in the seeds and stems of certain genera *(Ipomea, Rivea)* of morning glories. The psychoactive properties of this alkaloid is best known through its powerful cousin, the synthetic derivative, LSD-25 (d-lysergic acid diethylamide), produced by Albert Hofmann of Sandoz laboratories in Switzerland in 1938. LSD-25 is a mind-altering drug with a potency several thousand times greater than mescaline, and with a notoriety sufficient to require no further amplification at this point. It may be ingested, absorbed through the skin or mucous membranes, or injected. It is a potent hallucinogen that produces states described as heightened, intensified, or altered awareness of the world. This includes a number of auditory and visual phenomena, specifically emphasizing great heightening of the sense of color, and alterations in perception of space and time, among numerous other effects that produce a feeling of mental power, insight, and extraordinary perceptual sensitivity to one's surroundings. Naturally occurring lysergic acid amides may also be ingested or absorbed through mucous membranes and are also potent hallucinogens, albeit at significantly higher dose levels. While LSD-25 produces hallucinations at dose levels as low as 0.00002 grams, lysergic acid amides have an active dose about 100 times greater—circa $0.001^{-0}.002$ gram.[19]

The Mode of Preparation

We have been able to remove a few candidates from the list of drugs that might be (or have been) the original Soma plant simply because they are not native to India and could not have been imported there in Vedic times. These are the New World drugs peyote, mescaline, psilocybin, and cocaine. We have removed other drugs from consideration because their effects, though widely sought after, are not sufficiently intense—tobacco, coffee, and tea. We removed a further few—Sarcostemma, Ephedra, and others—because they are listed in texts as Soma substitutes if the real Soma cannot be found, and therefore cannot themselves be Soma. This leaves alcohol, cannabinols, opiates, ibotenic acid, and lysergic acid amides as those drugs that are both sufficiently potent to cause the effects recorded in the Rig Veda, and known to be present or available in India in Vedic times (ca 1000 B.C.).

In the strategy of elimination followed throughout most of the history of the quest for Soma, the next step has been fatal. Scholars have uniformly moved to compare the descriptions of the Soma plant and its substitutes, as they are given in the Vedas and the Brahmanas, with the plant sources of the drugs which remain in contention as possible "Somas." This strategy of procedure has led to a stalemate for the reasons given by O'Flaherty: those with enough Sanskrit to read the text had not enough botany to make an identification, and those with enough botany to make an identification had not enough Sanskrit to interpret the descriptions. The few remaining scholars who could handle both kinds of evidence repeatedly failed to make an unambiguous identification that could compel assent, for three good reasons. The first is that the best efforts at matching the morphology of existing candidate plants with the sup-

Natural Knowledge in Preclassical Antiquity

posed effects of Soma have been made by "amateurs" with no professional credentials either as Vedists or botanists, who could therefore be ignored with impunity even when they could not be refuted. The second reason is more creditable, but is also sociological rather than scientific. It is the reluctance of the majority of sober and capable scholars who have made a life-work of comparative religion and philology—a reluctance either religious or moralistic or simply aesthetic in origin—to see in the ancestry of a significant world religion an ecstatic experience induced by a hallucinogenic drug. The third and only scientifically sound reason has been that the texts by themselves are, from the standpoint of plant morphology, vague, contradictory, and ultimately inadequate for a definitive determination of the identity of Soma with any single known plant.

The impasse created by this confluence of reasons suggests that a new approach may be in order. Thus, rather than look for the plant, let us first examine the textual directions in the Vedic corpus and the Brahmanas for preparing Soma—for making the potent elixir. Then we shall be in a position to make some further reduction in our list, if it turns out that the active principle of some of the candidate plants on our list cannot be extracted in the manner prescribed.

Book 9 of the Rig Veda gives the directions for the preparation of Soma. The ritualist spread on the ground an ox or cow hide (9.65, 66), on which was placed either pressing stones (9.34) or a mortar (9.36). The Soma plants were then beaten or crushed to obtain a juice, which was either red, tawny, golden, or brown. Either before filtration (9.63, 103) or after, the Soma juice was blended with milk or curds. In either case, the Soma juice was poured through a sieve made of a woolen fleece, which cleaned the liquid of any fragments of the plants themselves ("The fleece retains his solid parts as though impure" 9.78.2). Soma is described in several places in the plural (see 9.64, 9.98), and many of the hymns in this book contain a rhythmic refrain "Flow Indu, flow for Indra's sake," which suggests that they are pressing songs—one would imagine the

The True Identity of Soma

pressing as hard work. But that's the extent of the preparation. The celebrant crushes the Soma, catches the liquid, filters it through a woolen sieve to catch any plant fragments, and then drinks it.

The Satapatha Brahmana gives much more complex and specific directions, which reflect the pharisaical elaboration of the original. The third Kanda of the Brahmana is devoted to the Soma sacrifice, the Agnishtoma, which begins with the complex reenactment of the purchase of the Soma plant from the Ghandarvas, semidivine beings who have appropriated Soma from the gods and become its guardians, or the guardians of its guardians. Soma is expensive, and comes from far away. A single sacrifice requires a price including gold, a cow, a goat, cloth, and a pair of kine (SB $3.2.4^{-3}.3.3$). The text specifies that Soma comes from the mountains, that the Soma be mountain grown. In this locus (3.3.4) we find statements such as "this surely means," or "in this there is nothing obscure," which are reminders that this is a commentary on works whose meaning is uncertain and obscure. There are references to Mitra, Varuna, and Surya that are taken directly from Rig Veda (book 10) interspersed as quotations and invocations in the text as, for example, SB 3.4.24, which repeats Rig Veda 10.37.1. At this point there is a great deal of interpolated consecrating ($3.4.3^{-2}1$), the function of which is to strengthen the Soma, a reminder that the potency of available Soma is on the wane.

But essentially the directions are the same. Pressing boards are laid down and a pressing skin of ox hide, dyed red, is laid on them. (Here as elsewhere we see that the original Soma may not be present. If you can't get the juice to run tawny or red [as the Vedas describe it], you can produce the correct appearance by pressing it out on a red hide.) The available Soma is divided into five portions, one for Indra, one for Gayatri, the falcon who brought Soma down from heaven, and the rest for other gods. The celebrant pours water over the Soma plants, then beats the Soma with a rock. When the plants are crushed, they are tossed in a bowl, where they soak up water, and then are wrung out before having new water added to them, whereupon they are pressed again. They are soaked, crushed, soaked, wrung out, soaked, wrung out again, and again, until they are used up—twenty-four pressings in all (SB 3.9.4⁻⁴.1.1.).

They are then consecrated again, with a plea that Soma, when he is slain (i.e., crushed) will be sweet and palatable and not putrid (4.1.1.13), and a plea to Soma that the minds of the celebrants will obtain him this time (4.1.1.22). At this point the gods' portions are set aside, and the celebrants begin on the remaining portions of Soma which, after pressing, and unlike the God's portions, are passed through a straining cloth. A series of libations are poured into the fire and earth, but by 4.3.3.1 the priests are quaffing the filtered Soma juice. That's it. Soaked in water, crushed, passed through a filter of wool, and drunk on the spot.

Now that we know this, what do we know? We know that in Vedic times, the plant was sometimes crushed when fresh—since the text does not specify that it should be soaked first. Contrarily, by the Brahmanic period, the Soma is purchased from someone who has gathered it far away, and it is therefore dry and must be soaked in order to yield any juice at all. We know as well that the active ingredient is water soluble, since it is wrung out with water. We also know that the solid parts of the Soma plants contain something one does not want to ingest, and therefore one filters the portions for human consumption. Only the gods drink their Soma unfiltered, and dire consequences await those who would attempt to drink the gods' portions. On the basis of these conclusions, we may move to reduce the number of plausible candidates for Soma.

The first candidate for elimination is the *bhang-ganja-charas* group obtained from the hemp plant *(Cannabis indica)*, the active ingredient of which is tetrahydrocannabinol. Though it has been proposed as a candidate for Soma by Mukherjee and Roy, it has been dismissed on the grounds that it contradicts the description of Soma and Soma substitutes (leafless, without flowers, resembling cows' udders) given in Rig Veda and the Satapatha Brahmana.[20] While this contradic-

The True Identity of Soma

tion (along with the judgment that Mukherjee's argument consists mostly of a combination of "linguistic reasoning with the purest twaddle"[21]) would seem to exclude the hemp plant categorically, the fact that we do not have a definitive description of the Soma plant has allowed the "bhang theory" to be offered again and again.

A more forceful exclusion is obtained with reference to the manner in which Soma is prepared, namely, by *solution* (in water or water mixed with milk or curds) and filtration. Cannabinols are insoluble in water. No matter how much you crush and soak and wring and crush again, you cannot mechanically make a marijuana plant give up its psychoactive ingredients. The plant must be ingested directly or smoked. True, it grows in India; true, it has a Sanskrit name; true, it is intoxicating and may even be classed as a hallucinogen, but it cannot be the Soma plant as prepared according to the directions in the Rig Veda and the Satapatha Brahmana. The Soma solution is filtered before it is drunk, specifically to remove plant fragments that might otherwise be imbibed, and thus even the improbable and remote chance of intoxication by consuming fragmentary pieces of the leaves or resin is removed.

The next plant to be eliminated by this route is opium *(Papaver somniferum)*. Opium is gathered by slitting the unripe capsule (a bulb atop the stem and directly below the flower) of the poppy plant. This capsule when slit oozes a gummy latex that, after drying, is scraped and worked into a cake, portions of which may be eaten, smoked, or imbibed in an alcoholic tincture. Opium contains several alkaloids with medical applications, including papaverine and coedine, but the principal narcotic alkaloid is morphine. Morphine cannot be extracted from raw opium by solution in water.[22] Since opium is often prepared for smoking or for the making of tinctures by grinding the dried cake to a fine powder, it is conceivable that it could be drunk suspended in aqueous solution, even though filtration would remove all but the finest particles; but there is no record in the whole literature of opium use (and a compendious literature it is) of this approach to opium ingestion. More-

Natural Knowledge in Preclassical Antiquity

over, the directions for preparation in the Rig Veda and the Brahmanas speak of the bruising of the plant. The sap of the opium plant (as opposed to the latex of its fruit capsule) contains oxalic acid present as microscopic insoluble crystals.[23] Oxalic acid, present in rhubarb leaves and a number of common house plants, causes rapid swelling of the mucous membranes of the mouth and throat, followed by laryngospasm (inability to speak—thus the name "Dumb Cane" for a favored plant often observed in lobbies and waiting rooms) and death by suffocation. Attempts to produce an effective dose of (insoluble) morphine by means of the Vedic preparation of Soma would produce unacceptable concentrations of this acid. While opium has rarely been proposed as a candidate for Soma and might be eliminated on other grounds, this approach to the question again definitively excludes it.

With the alcohol theory out of the question—the immediacy of pressing to ingestion excludes fermentation, which takes days, weeks, or even months—this leaves only three categories of hypotheses concerning Soma/Haoma. One, Soma/Haoma is a rather mild herbal concoction (ephedra, milkweed, horsetails) the effects of which have been wildly exaggerated in the surviving texts. Two, Soma/Haoma is a now-extinct or rare and in any case unknown plant with potent hallucinogenic properties. Three, Soma/Haoma is the name for an aqueous filtrate of known plants of hallucinogenic potency that yet deserve our consideration.

The first hypothesis may be eliminated on the grounds that it denies what the texts say, and invokes implicitly a set of subsidiary psychological and anthropological hypotheses concerning the misinterpretation and misrecording of a long-standing tradition by its recorders (translation: either they didn't know what they were talking about or they were lying). This is absolutely unacceptable, as it is an attack on all documentary evidence. It does not identify Soma, but explains it away as a delusion or a deliberate fiction.

The second hypothesis, that the real Soma is as yet undiscovered, has inherent plausibility, but logically it must be re-

The True Identity of Soma

served until all the remaining candidate plants are demonstrated to be neither potent, nor present in India at the
appropriate places and times, nor preparable by the instructions we have; and that none of them fit the descriptions of or
tropes for Soma/Haoma.

On the hypothesis that Soma does exist and is hallucinogen-
ically potent, and is neither extinct nor nor the product of some
extant but unknown plant, our list of candidates is reduced to
two: the ibotenic acid/muscimole alkaloids from fresh and
dried *A. muscaria* mushrooms, and the lysergic acid amide
alkaloids from grasses and domesticated grains infected with
Claviceps fungus (ergot).

The hypothesis that the Soma plant is indeed the fly agaric
mushroom has been developed with great care and skill by
the ethnomycologist R. Gordon Wasson in his *Soma: Divine
Mushroom of Immortality. Amanita muscaria* is hallucinogenic and potent. It produces the effects described for
Soma—power, strength, clarity, elation. It grows in the mountains, and Rig Veda 9 asserts over and over again that Soma is
a mountain plant. Its juice is golden yellow, and the most
common epithet for Soma, as Wasson has made clear, is *hari*
[Sanskrit cognate with *hiranya,* golden]—a bright, flaming yellow, yellow-brown, or chestnut.[24] Its name (Fly Agaric) explains to a certain extent the extremely puzzling (otherwise)
passage in Rig Veda 1.119: "To you, O Asvins, that fly betrayed
the Soma." Wasson conducted experiments in which he split
the stipes (stems) of Fly Agarics and let them bleed. The juice
attracted flies, which drank the liquid and collapsed, and later
revived (p. 61). The dried mushroom (unlike the fresh) contains muscimole, an alkaloid about ten times as potent as ibotenic acid; muscimole is readily soluble in water, so that an
aqueous filtrate of dried, soaked, crushed *A. muscaria* would
produce a powerfully psychoactive solution on the spot. It
fulfills every criterion we have established for Soma—provenance, effect, mode of preparation, color, tropes, even the "be-
trayal of the fly"—every criterion but one. If Soma is *A. muscaria,* then there is absolutely no reason to filter the Soma

through a fleece of wool. One can eat fly agaric mushrooms, dried and fresh. In Wasson's painstaking accumulation of European accounts of Siberian *A. muscaria* ingestion, there are reports of dried mushrooms being boiled and the liquid drunk, but the majority of accounts are of eating—chewing or swallowing whole—with quantities of water and sometimes dilute vodka. When Wasson himself attempted to produce *Amanita* intoxication by ingestion of *A. muscaria* in Japan in 1965-6, he ate the mushrooms. He obtained no intoxicating effect, by the way. The only member of the experiment who experienced the ethnologically documented effects of *Amanita* was a Mr. Imazeki, who in one of the apparently plural trials toasted (and therefore dried) his mushrooms first, and experienced a tremendous elation. Wasson concluded that "drying is of the essence" (p. 155). That may be, and pharmacologically it is indeed expedient so to do, and this may be of interest in more general investigations of ethnomycology. But the present strategy in pursuing this inquiry into Soma has been to assume that each of the steps of the original Vedic technique of preparation of Soma was necessary, and that all taken together could uniquely identify the plant. No matter how baroque the additions to the Soma ceremony in the Brahmanas, the essential steps in preparation were three: soaking, pounding, and filtration. It must be that filtration had some essential function and practical foundation concerned with the characteristics of the plant itself.

Wasson himself was cognizant of this problem: "Early in my inquiries I came upon a verse, 9.156, that seemed to present an obstacle to my fly-agaric thesis. According to the poet of this hymn, when the Soma plant is pressed and then run through the woolen filtre, it leaves behind in the filtre its 'knots' or 'nodules.' (The Vedic word, in the plural, is *parusa*.) There has been some difficulty with the sense of the sentence, but agreement on this particular word. Now mushrooms have no knots or nodules, which are characteristic of shrubs and trees. Here was a hurdle to cross or I was in trouble" (p. 59). Indeed, the leaving behind of the solid parts of the Soma plants is

The True Identity of Soma

mentioned later in Rig Veda 9.78.2: "The fleece retains his
solid parts as though impure." Wasson explained his way out
of this dilemma to his own apparent satisfaction by identifying
the "knots" in question as the scales and patches of white on
the red cap of the fly agaric, citing as his inspiration S. S.
Bhawe's 1957 commentary on this verse, which discussed the
description of the knots as sticking to Soma's body, and as well
their "shining" character.

This seems neither sufficient nor necessary. The little white
patches on the head of the mushroom are just that. They are
not islands of toxicity that must be removed, and the cumbersome and lengthy process of filtration could hardly be justified
solely as a cosmetic operation to remove little white patches
even if, after a thorough bashing on stones, there were any
little white patches left intact, which is in itself quite doubtful.
Rather, it seems to me that the absence of any knots or nodules
on Fly Agarics to be caught in a filter is reason enough, if not
to dismiss A. muscaria entirely, at least to go on looking for
another plant that fits the criteria, and for which filtration of
the aqueous solution is absolutely necessary.

This brings us to our final candidate, lysergic acid amides
from grain and grass infected with Claviceps (ergot) fungus.[25]
Ergot is the name for the fungus and the disease it causes. The
generic name Claviceps means "club-headed." It is parasitic
on wild grasses and most cultivated grains, on hay and pasture
grasses, sugarcane, and even bamboo.[26]

All of the Claviceps species have a similar life history on all
their hosts. The fungus winters over in a hard sclerotum—a
dense mass of fungal cells which has a different shape for each
species—oval, elongated, even spherical—with a black or purple exterior, slightly furrowed.[27] Sclerota must undergo a period of dormancy lasting from several weeks to several months,
and often requiring a period of low temperature. They germinate in the spring after their host (grain) has flowered (formed
a head). Then each sclerotum sends out tiny mushroomlike
stalks each with a spherical head (hence Claviceps). Beneath
the head are thousands of spores, which are then expelled to

Natural Knowledge in Preclassical Antiquity

be carried by the wind, or which ooze out to be spread by flies.[28] Those spores that find their way to the host plant germinate and produce a mycelium, which replaces the ovary of the flower, thus aborting it. The mycelium, within several days, produces millions of spores called conidia, along with a sweet liquid called honeydew.

"The honeydew may be produced in such quantities that it oozes out between the glumes, laden with conidia. This conidia laden honeydew attracts insects, especially minute flies that visit grasses for their own purposes, and, if the weather is humid and the temperature right, the fungus is spread rapidly to other flowers of the same plant, and the flowers of nearby plants. . . Given favorable conditions in the way of moisture and temperature, an epidemic can result."[29]

The formation of sclerota, the horn-shaped pegs that project from the ripening ears of grain, follows the production of the conidia by the mycelium, and completes the life cycle of the fungus.

Ergot has been of chemical and pharmacological interest for a long time. Its most interesting chemical constituents from the pharmacological point of view are its alkaloids, a term I have used in this essay but have not defined. Alkaloids are alkaline organic chemicals—colorless crystalline solids that occur in nature dissolved in plant saps and that combine with common organic acids containing carbon, hydrogen, and nitrogen. They are generally soluble in water or acids, and of all the "natural products," they produce the most pharmacologically interesting and potent compounds: physostigmine, curare, strychnine, ergotamine, nicotine, caffeine, morphine, quinine, cocaine, psilocybin, lysergic acid, and reserpine being among the more notable medicinal alkaloids.[30]

Ergot contains more than forty individual alkaloids, of which about two dozen are of the important class of *indole* alkaloids. While the first indole alkaloids of ergot were isolated in crystalline form in the late nineteenth century, their chemical structure—and their common chemical nucleus—was identified only in the 1930s and named lysergic acid.[31]

The True Identity of Soma

The first medicinal product of this research came from laboratories in Switzerland. A purified alkaloid breakdown product of Ergotamine, called Ergobasine (brand-named Methergine and Oxytocin) was developed and is still employed in obstetrics. It hurries labor by encouraging uterine contractions, and is also a hemostatic (it stops bleeding) widely in use to stop postpartum bleeding.

In an extension of this research, Albert Hofmann in 1938 produced lysergic aid diethylamide, abbreviated LSD-25 (for Lyserg-Saure-Diethylamid, "25" because it was the twenty-fifth lysergic acid derivative in the series), while searching for compounds that were not "uterotonic" but that might have important properties.[32] By his own account its properties were not impressive, and it was not until 1943 that he again synthesized the compound and had the following experience, which he reported to his laboratory superior.

> Last Friday, April 16, l943, I was forced to interrupt my work in the laboratory in the middle of the afternoon and proceed home, being affected with a remarkable restlessness combined with a slight dizziness. At home I lay down and sank into a not unpleasant intoxicatedlike [sic] condition, characterized by an extremely stimulated condition. In a dreamlike state with eyes closed ... I perceived an uninterrupted stream of fantastic pictures, extraordinary shapes with intense kaleidoscopic play of colors." (p. 15)

While Hofmann later experimented extensively with LSD ingestion at Sandoz, in this first instance he absorbed it accidentally through his skin. Thus began the history of the best known of the hallucinogens. Hofmann's interest in the chemistry of hallucinogens brought him into contact with Gordon Wasson, who was at that time (middle 1950s) investigating Mexican shamanistic cults based on ingestion of mushrooms, and this collaboration eventually led (1958) to Hofmann's isolation and naming of the active ingredient: "psilocybin" (p. 118). Finally, again in collaboration with Wasson and with the help of research by Richard Schultes, director of the Harvard

Natural Knowledge in Preclassical Antiquity

Botanical Museum, Hofmann began a search for the active ingredient in a hallucinogen, called *ololiuqui,* employed by another Mesoamerican Indian cult. This drug was prepared from the seeds of morning glories of the species *Rivea corymbosa* and *Ipomea violacea.* The manner of the preparation of the drug was described by Wasson in 1963. The natives prepared the seeds for ingestion by grinding them on a grinding stone and then soaking them in water, after which the mixture was passed through a cloth strainer and ingested.[33] In 1959-60 Hofmann and his assistant Hans Tscherter determined that the active ingredients in *ololiuqui* were lysergic acid amides, which had never before been observed in higher plants (i.e., vascular plants).[34]

Investigations of ergot in the 1930s, 1940s, and 1950s in Europe were concentrated on the species *Claviceps purpurea,* easily and locally obtained. *Claviceps purpurea* has not been observed to produce free lysergic acid, nor to accumulate lysergic acid amide (ergine). Therefore it was long assumed that lysergic acid amide could only be obtained as a hydrolysis product (chemically extracted in the lab) in significant amounts. But in the 1960s it became clear that another species of ergot, *Claviceps palspali,* which infects the grass *Paspalum distichum* (knotgrass, a common roadside weed throughout Eurasia and North America) does produce and accumulate ergine—lysergic acid amide—in substantial quantities, so that the hallucinogenic compound is available naturally after all in ergot-infected plants.[35]

If we are to surmise that the active ingredient in Soma is lysergic acid amide from ergot sclerota on grasses and grains, then we must establish that the grains employed as hosts by ergot were or are available in India. An easy checklist is whether there is a Sanskrit name for the plant. This immediately eliminates one candidate—*Lolium tementulum* or Darnel, which infects with *C. purpurea,* and for which there is no Sanskrit name.

There are, as it turns out, still a number of candidates. There is the foxtail millet (*Setaira italica*), in Sanskrit *Syamaka,* a

The True Identity of Soma

tawny to red-brown cereal grain whose panicum ("ear," the seed clusters at the top, as opposed to the stem) can be tens of centimeters long and up to three centimeters in diameter. There is barley (*Hordeum vulgare*), a golden grain that infects with *Claviceps purpurea*, and is called *Yava* in Sanskrit. There is broomcorn millet (*Panicum milaceum*), Sanskrit *China*, a golden to red-brown grain with long nodding panicles. There is *Paspalum scrobiculatum*, Sanskrit *Kodrava*, a cereal grain native to India, which infects with *Claviceps paspali*. Finally there is finger millet (Indian millet), *Eleusine coracana*, called *Ragi* (a Kanarese word) and in Sanskrit, *Soma*.

The next step would be to establish whether they were in India in Vedic times. *Eleusine coracana* appears in the archaeological record in India in 1800 B.C., *Paspalum scrobiculatum* around 1200.[36] The Indus Province reveals wheats and a barley, *Hordeum vulgare*, at early levels.[37] *Panicum milaceum* and *Setaira Italica*, two of our candidate grains as hosts for Soma, are among the oldest cultivated grains—they are beer grains—and were in cultivation in Europe and China in 5000 B.C.[38] They are today economically important grains in India and may be presumed to be early arrivals, since "a number of the principal grains and plants found in the Indus Province and even in adjacent parts of the peninsula, appear to have been first domesticated in West Asia and to have entered South Asia probably along with the spread of cultivation in subsequent centuries."[39] None of the words for specific varieties of grains cultivated in India is of either Indo-European or Indo-Iranian origin. Both Indo-European and Aryan contain only words for "grain-in-general."[40] Since the Vedas are derived from the culture of the Indo-Aryans or Aryans who arrived in northwestern India in the middle of the second millennium B.C., it is notable that of the very small fraction of modern agricultural terms in Hindi derived from Aryan roots (3.5 percent), all the terms that provide equipment and raw materials for the Soma ceremony are Aryan in origin: "grain," "hide," "wool," "cow," "curds," and "reap."[41] All of the candidate grains tolerate the climate of northwestern India and the Himalayas:

Natural Knowledge in Preclassical Antiquity

wheat and barley are derived from wild steppe grasses; the millets are tolerant of altitude; *Eleusine coracana*, finger millet, "Soma" is cultivated in the Himalayas and along the Afghan border up to about 8,000 feet.[42]

This is important because Soma must be mountain-grown to fulfill the original stipulation of the Rig Veda. But it is important for another reason. Plants producing alkaloids have long been known to occur as chemical races.

> A chemical race is an intraspecific variety distinguished from other plants of the same species only by its chemical composition. The intensive breeding of medicinal plants for high content of some active constituent has produced some of these races, but others appear to have arisen naturally and to have persisted in geographical isolation . . . formation of alkaloids varies notably from tissue to tissue within the same plant, and also changes during the course of ontogeny. This kind of variability as well as the occurrence of chemical races can account for disagreements over the presence or absence of a particular alkaloid in a given plant.[43]

This striking property is much in evidence in ergot species. "The compounds of interest and importance in ergot occur in different amounts in sclerota gathered from a single host species growing in different places in the same season, in sclerota from different species of plants growing in the same area in the same season, and in sclerota from the same species of plant growing in the same place in different years."[44] Clyde Christensen has made the analogy with wine—the same vinyards producing different qualities of wine grapes in different years, and different vineyards producing different qualities in the same years.[45]

There are two clues as to why mountain-grown Soma would be preferable. The first is the fact that the sclerota of *Claviceps* species require a period of cold-weather dormancy as a part of their life cycle, and this is more regularly obtained in mountainous areas. The second is that long photoperiods and exposure to ultraviolet light concentrate alkaloids in plant tissues.

The True Identity of Soma

There is conspicuously more ultraviolet light at altitude than at sea level. For purely chemical and biological reasons, potent chemical races of ergot would be found at altitude rather than lower down.[46]

We have now determined that ergot species produce in grains the substance lysergic acid amide, which is a hallucinogen and is the basis of a Mesoamerican drug cult in its preparation from morning glory seeds. We have documented that grains infected with ergot are cultivated in India, and that they were cultivated there in Vedic times. We have seen that mountain-gathered Soma is likely to have higher concentrations of lysergic acid amide (and other alkaloids), because of the survival of chemical races which are exposed to long photoperiods and more ultraviolet radiation.

We have now to ask whether the mode of preparation, the tropes, and the descriptions of Soma garnered from the Rig Veda fit grains infected with ergot. Since the only other candidate hypothesis for Soma still under consideration is Wasson's *A. muscaria*, let us see how the selection of descriptions that Wasson assembled concerning Soma (to make a case for Soma as *A. muscaria*) fit the assumption that Soma is lysergic acid amide obtained from preparations of ergot. The reader may judge whether the descriptions fit this hypothesis of ergot-infected grain better than that of a red mushroom with a white stalk.

1. "The fathers with a commanding glance laid the germ." (Rig Veda 9.83.3)

It is clear that Soma is parasitic on and not a part of the grain host. Thus the ontogeny of the Soma's growth in the grain is seen to begin when the gods "lay" the germ of the Soma in the grain.

2. There is no reference to root leaf or blossoms.

A negative doesn't tell us much. Mushrooms don't have root leaves or blossoms, but neither does a sheaf of cut grain—and we harvest grain with a reaping hook, we do not uproot it.

Natural Knowledge in Preclassical Antiquity

3. It comes from the mountains. (Rig Veda 9.82, 85, 87 ff.)

Amanitas are only found in high mountains in India.[47] Wasson asserts that *Amanita* only grows in birch and coniferous forests, but this is not always the case, even with old world varieties.[48] No one has ever established that there is an altitude-correlated chemical raciation of *Amanita,* thus there is no pharmacologic advantage to mountain-grown plants, as there is with *Claviceps* species. But in India, if *A. muscaria* were Soma, it would have to come from the mountains.

4. The Soma plant is pressed according to directions we have already reviewed—pressed, soaked in water and wrung out, pressed again.

Both *Amanita* and *Claviceps* can be prepared this way. Both ibotenic acid and lysergic acid amide are water soluble. Again, both plants are candidates on this count.

5. "The atharvins have mixed milk with thy sweetness." (Rig Veda 9.11.2.)

Amanita muscaria tastes terrible, so terrible that it is usually swallowed in chunks without chewing, and washed down with cold water. *Claviceps*-infected plants, on the other hand, are covered with honeydew, which tastes sweet. Wasson is interested in the passage because of the blending with milk or curds, and does not take up the issue of the terrible taste of *A. muscaria,* and the obvious disjunction of this with tropes of Soma's honeylike sweetness, which are ubiquitous. Wasson offers no explanation of why *A. muscaria* would be blended with milk or curds, although there is an excellent biochemical reason for doing it with *Claviceps,* to be explained when we get to filtration.

6. The most common epithet for Soma is Sanskrit *hari.*

Hari means yellow. The *A. muscaria* is red and white. Ripe grain is golden yellow, as is the honeydew bearing the conidia of the fungus. All the grains in question are tawny yellow to

The True Identity of Soma

red-brown, but most are what botanists call "straminous"—the golden color of straw.

> 7. Soma is like an udder to be milked. (Rig Veda 3.48.3 and 8.9.19. "The one with the good hands has milked the mountain-grown sap of the sweet honey, the amsu has yielded the dazzling drop." (Rig Veda 5.43.4.)

One could see the stipe of an *Amanita* as an udder, but the panicles of ripe infected grain also would look like an udder, and are actually covered with sweet honey-colored drops that can be stripped off.

> 8. Soma is always filtered. It passes through three filters. The first is the filtering by sunlight: "he who is cleansed by the sun's ray" (Rig Veda 9.76.4); the second filter is the woolen fleece, which is mentioned everywhere the filtering process is discussed at all, both in the Rig Veda and in the Brahmanas; the third filter is the heart of the recipient (Rig Veda 9.73.8).

The Soma plant is sprinkled, crushed, wrung out, and filtered through a fleece. I have suggested before that there is no reason to filter *A. muscaria,* because the normal mode of ingestion is swallowing, or drinking a boiled infusion of dried mushrooms. In other words, if you eat *A. muscaria,* nothing untoward happens. We may then ask what happens if you were to eat ergot sclerota, rather than drinking the aqueous solution filtered through wool. The answer is that if you were to ingest ergot sclerota, you would die. Ergot contains a virulent poison called Ergotamine (or Ergotoxine). When infected grain is harvested and ground into flour and ingested, as happened with depressing frequency in Medieval and Renaissance Europe, and as late as 1927 in the USSR, the unfortunates who eat bread made from infected flour develop a disease called ergotism, also known as "St. Anthony's Fire." It has two forms, convulsive and gangrenous. In convulsive ergotism, which may drag on for months, one experiences violent and terribly painful convulsions, blindness, deafness, diarrhea, and the opening of watery pustules on the extremities. Later, one dies.

Natural Knowledge in Preclassical Antiquity

The gangrenous form begins with lumbar pain and limb pain and swelling, followed by violent burning pain (St. Anthony's Fire), alternation of intense heat and cold, followed by gangrenous attrition and blackening of limbs, and emaciation. Then one's arms and legs fall off. Then one dies.[49]

This is why one blends Soma juice with milk and curds, why one filters it, and why the filter must be of wool. All three are requisite. The lysergic acid amides which produce the hallucinogenic intoxication for which Soma was sought are water soluble, and in fact "more soluble in water than the other principal alkaloids of ergot."[50] Ergotamine, the alkaloid that contains the poison, is almost insoluble in water but is soluble in fat. Since in the pressing and wringing one is extracting the lysergic acid amide in aqueous solution, the filtration is necessary lest one ingest the Ergotamine in the form of fragments of the ergot sclerotum. One filters Soma because the solid parts are poison.

Why should wool be specified for the filter? There is a good reason on chemical grounds, connected with the blending with milk or curds. Since some slight amount of Ergotamine will have gone into water solution, mixing with milk will offer the Ergotamine fats into which it will readily dissolve. One then pours this combination through the filter, and employs the tremendous fat-absorbing power of sheep's wool to retain the fats that now are bound to Ergotamine. There is even the additional possibility that the enzymatic action—whether of milk or bacteria in milk—would make even more lysergic acid amide. Finally, the structure of moist wool takes a molecular configuration called a "Beta sheet," which acts as a molecular sieve, able to catch molecules of certain shape, especially those with large "side chains" which get caught in the molecular filter. This is, however, not demonstrated in this case. The point is rather that Soma must be filtered to prevent ergotism, and a wool filter is the best kind there is for this purpose, because of its ability to absorb the fat-dissolved Ergotamine, without losing the lysergic acid amides. There is no similar rationale for *A. muscaria.*

The True Identity of Soma

These properties seem to me to be decisive in favor of *Claviceps* as the source of Soma, and I think it is clear that in other crucial ways—color, sweetness, tropes—this is a much better explanation, since it explains every aspect of the Soma preparation and explains nothing *away*.

Note that there are three filtrations, the middle of which is the sheep's fleece. The first filtration is that performed by the sun's rays (Rig Veda 9.76.4); this is an acknowledgment of the role of long photoperiods (and therefore increased ultraviolet radiation) in fixing Soma in the grain. The final filter is the heart of the recipient/percipient (Rig Veda 9.73.8), which formula recognizes the distinct character of this sort of hallucinogenic experience when contrasted with alcohol intoxication—the heart knows.

Thus far we have concentrated on a specific strategy: essentially biochemical and pharmacological rather than descriptively botanical. We have also concentrated on Vedic description for the most part, rather than Brahmanic glosses. Finally, we have not had much to say about the Soma substitutes, which seem to have led so many Vedists and botanists so far astray. However, in the context of discourse we have established, we can now turn to the identity of the Soma substitutes, and show how their confusing variety actually serves to emphasize the identity of Soma as expounded herein.

The Satapatha Brahmana sets out the order of the Soma substitutes (SB 4.5.10.2⁻⁶). The first is the "reddish brown" *phalguna* plant, which is "similar to Soma." The observants are enjoined not to use the bright red *phalguna* plant. The *phalguna* was identified by Wilson in 1832 as *Arjunata pentaptera* (now *Terminalis arjuna*), a myrobalan—a red tree growing in the Himalayan foothills whose bark is a mild cardiac stimulant and diuretic.[51] This misidentification is the impermissible "bright-red" form. The actual Soma substitute is the Sanskrit *Arjuna* plant, *Lagerstroemia Flos-Regina,* a shrub whose crushed seeds are narcotic.[52] It is substituted because it works like Soma, not because it looks like Soma. The next two substitutes, the *syenahrta* plant and the *adara* plant, still elude identification. The next two substitutes thereafter are grasses: brown *durva* grass (*Eragostris cynosuroides*) and yellow *kusa* grass (another species of genus *Eragostris*). These are clearly chosen because they look like Soma.

Among other substitutes is a plant that actually figures as a candidate for Soma—*Syamaka* millet (*Setaria italica*, foxtail millet), described in the Satapatha Brahmana as "among plants doubtless [is] most manifestly Soma's own."[53] There is a lovely poignance in this description. The plant is one of the right plants, but of course as a plant (not infected with ergot) it produces no effect, and is therefore not a major substitute for Soma.

The odd occurence of creepers *(valli)* in the later substitutes for Soma makes better sense once the biochemical foundation of Soma as the name of a principle of action rather than the name of a specific plant is understood. For instance, there is the case of Somaraji, a favorite Soma candidate among the early European Vedists. Monier-Williams incorrectly identified as *Somaraji* the plant *Somaraja* (*Vernonia Anthelmintica*—purple fleabane), a potent vermicide. The confusion is based on the nominal similarity between the Sanskrit *Somaraji* (*Psoralea corylifolia*—Babchi) which is also a vermicide, and the later Hindi *Somaraji* (*Paederia foetida*—moon creeper). The latter plant is a member of the *Convolvulaceae*—morning glories, and may therefore contain in its seeds lysergic acid amides.

All of the acceptable Soma substitutes either look like Soma—that is, they are straminous to reddish brown grasses—or they have pronounced narcotic properties. The terrible confusion of the two in the *Brahmanas* is a consequence of the loss of the original sources of the plant, and the subsequent incomprehensibility that the ones that look like Soma do not act like Soma, and the converse.

What then of the identity of the original Soma? The question is a trap. Once we have identified Soma as a principle of action rather than a specific plant, it could have been several. My preferred candidates are *Eleusine coracana* (finger mil-

The True Identity of Soma

let—Sanskrit *Soma*) and *Paspalum scrobiculatum* (*Kodrava*). My preference for the former is in its name, its status as a host for claviceps species, and because E. B. Havell identified it on morphological grounds as the original Soma in 1920, based on his examination of the evidence collected by the Afghan Boundary Survey.[54] My preference for the latter is that it is native to India, that it is host for *Claviceps paspali*, which produces free lysergic acid amides, and the following description from Nadkarni's *Indian Materia Medica:* "The new grain is said to be powerfully narcotic and is eaten only by the poor who prepare it in various ways and from use are able to use it with impunity."[55] I take the latter phrase to mean the avoidance of Ergotism.

I think that it ought by now to be clear that Vedic Soma was a potent hallucinogen, available from grains infected by ergot fungus, and safe when prepared in the prescribed manner; Soma was a god, and he resided in the plants as Agni resided in fire. We need not know how Soma was lost to identify the living principle.

Nothing in my identification is designed to cast doubt on the authenticity of mystical experience, either chemically induced or self-made. It seems to me rather an argument that here, as in the other essays in this book, we may understand certain aspects of mythic traditions only through the addition of information garnered from modern natural sciences to the laboriously gained insights provided by philology and cultural history. It seems to me that the character of the experience induced by lysergic acid, available in hundreds of accounts from the last few decades, accords quite well with the profound insight into the meaning of life and death that Vedic celebrants sought through the ritual ingestion of Soma, and helps us to understand the powerful immediacy of these experiences as recorded in the Rig Veda, in an age when most of our religious experiences—insofar as they are ritual ingestions—have been reduced to polite, if earnest, commemorations.

7

Plato's Myths

The physiology of the senses is a border land in which the two great divisions of human knowledge, natural and mental science, encroach on one another's domain; in which problems arise which are important for both, and which only the combined labor of both can solve. *H. von Helmholtz, 1867*

For anyone who harbors a theory of myth in which myth and ritual practice or performance are entwined, the teaching of medicine in a modern university hospital provides a number of useful examples. First-year students in medical schools dissect a human corpse. In so doing they learn anatomy, they recapitulate the foundation of modern allopathic medicine in an understanding of this anatomy, and, willingly or not, they meditate on death. In their clinical years, the third and fourth years of medical training, they participate in "rounds" and, less frequently today, "grand rounds." In grand rounds, residents, interns, and medical students, under the watchful and often wrathful eye of the chief of medicine, debate at bedside the diagnosis, prognosis, and treatment of patients under their care. On graduation they take the Hippocratic oath and swear, among other things, to do no harm. They adopt the caduceus as their symbol—the winged staff of Hermes, entwined with serpents.

These are rites of passage. But they endure for the physician throughout life in the ceremony of the CPC, the clinico-pathological conference. The proceedings of one of these ceremonials can be read in any issue of the *New England Journal of Medicine,* under the heading "Case Records of the Massachusetts General Hospital." But the reading is only the pale

Plato's Myths

shadow of the real event, in which the medical staff, residents, interns, and students gather in an auditorium. Today this lacks some of the drama the ceremony had in the last century, when such conferences were held in an operating theatre—a true amphitheatre on the classical Greek model, with a semicircular arena surrounded by steeply raked rows of seats, allowing students to look down onto surgical operations in progress. But the ceremony proceeds today nonetheless, even in padded seats facing a proscenium stage, or on folding chairs in a lino-leum-floored hospital conference room under the glare of fluo-rescent lights.

The panel assembles on stage. The presenter steps forward and presents the case. The formula is always the same: A patient of such and such an age and description was admitted to the hospital complaining of certain symptoms. Various tests were made, physiological values ascertained, therapies instituted. Further problems, difficulties, and complications ensued, and counter- and additional measures, medical and/or surgical, were advanced. Eventually, there was an outcome, not always reassuring, as in many of the most complicated and interesting cases, the outcome is death. Then comes the question: what is the diagnosis?

Then the members of the audience begin their dialogue and debate. They retell the case in their own words, repeating from memory; they compare this case with others they have seen, they rehearse the evolution of the case, the living dialectic of the treatments and the patient's responses to them and finally come to a logical closure and offer their diagnoses. Arguments, often heated, ensue.

Then the denouement. The pathologist, who has studied microscopic samples of diseased tissue from the patient, steps forward and gives his diagnosis. He has not seen the patient, but he has seen the disease, and he shows pictures of his micro-scopic slides of the patient's tissues to the audience. His verdict is final. The dramatic resolution is, of course, in whether his verdict confirms or denies that of the presenters and audi-ence—and the possibility that it will not be confirmed is the point of the exercise, which pits the dialectical, logical, and communal practice of the guild of physicians against their accumulated store of book learning and medical science. By regular attendance at such conferences (or by the less satisfactory but sometimes necessary expedient of reading about them), the physicians develop their art. They do so constantly anyway, in the hospital locker room, in the hallway, in the dining room. Introduce two physicians and if their specialities coincide, and one can step aside far enough to leave them alone but still eavesdrop, one hears some variation on this conversational opener: "I had an interesting case the other day, let's see what you make of it." The "presenter" will then give the case in the form of the CPC: the symptoms, the treatments, the complications, the outcome, and the other physician, "the audience," is asked to guess the diagnosis. Then the exercise reverses and the audience becomes the presenter, and so on.

The real, live, ceremonial clinico-pathological conference provides a vital affirmation of the life of the community, in which one must practice the complexities of differential diagnosis, understand the logical meaning of the laboratory values representing the various aspects of the patient's physiology, and bring these quickly and decisively to a conclusion, which can then be enacted as a treatment. The discussion of the presenter and audience affirms, in performance, the art of medical practice.

This affirmation of art in a community and its contrast with written science leads us to a consideration of Plato's myths. This is the last topically substantive essay in a collection devoted to the study of natural knowledge in preclassical antiquity. With Plato we are almost entirely over the horizon of *mythos* and into other forms of holding natural knowledge within the Greek world—*historia, physiologia, philosophia.* Within the generations either side of Plato's, the major lines of Greek science were laid down: the medicine of Hippocrates; the physics, metaphysics, and natural history of Aristotle; the astronomy of Aristarchus; the geometry and optics of

Euclid; the botany of Theophrastus; and the geography and geodesy of Eratosthenes.

Plato sought in many and perhaps all of his dialogues to solve the "problem of knowledge" under emergency conditions. The problem of knowledge is not what we know, but how we know that what we know is true. In Socrates' lifetime, and partly through Socrates' efforts, the mythic order's integrity was broken. Since a myth is a story, when we ask: "how do we know it?," our response is that "we heard it," or "so it is told." If, however, our interlocutor then asks us how we know that what we heard was true, we are subject to all the emergent problems of the age of Plato and Socrates. That is to say, the corrosive effects of literacy, dialectic argument, conceptual thinking, historical comparison, and written communication upon the integrity of the moral order, formerly maintained by myth, were everywhere apparent to Plato, as they were to Socrates before him. Mythic knowledge is fundamentally uncritical; it is in this respect cognitively allied to common sense. It is the peculiarity of common sense knowledge that when critically challenged, it cannot defend itself by explaining its ground and still maintain its authority.

An example of the new attitude toward knowledge and therefore toward myth can be seen in the dialogue *Phaedrus* (about which we shall have more to say). Phaedrus and Socrates discuss the myth of Oreithyia—an abduction of a maiden by the North Wind, which supposedly happened near where they were walking. Socrates says he is willing to join the pundits in rejecting the myth as a true tale, and in rationalizing the myth as a euhemerism, a story of the fate of mortals embroidered to make it an encounter with gods. Socrates then says, however, that he does not envy the fate of the rationalizer: forced to ever more ingenious and labored explanations as he moves from simple abductions of girls by winds to stories of monsters like Chimaera (head of a lion, tail of a snake, body of a goat), "not to mention a whole host of such creatures, Gorgons and pegasuses and countless other remarkable monsters

Natural Knowledge in Preclassical Antiquity

of legend flocking in on them."[1] Socrates tells Phaedrus that it will take a lot of leisure to force all these into plausible shapes by rough ingenuity. He would rather follow the Delphic injunction and know himself—to find out what sort of monster he is. Socrates says that it will force the limits of his own powers to make himself into a being comprehensible by himself and, therefore "I can't as yet 'know myself'. . . and so long as that ignorance remains it seems to me ridiculous to inquire into extraneous matters" (230a).

Plato, as the prime successor to a radical and revolutionary thinker, inherited the problems disclosed by Socrates' approach to this emergent difficulty in our knowledge of the order of things. Socrates proceeded by dialectical inquiry into meaning, by the critical demolition of pretenses to knowledge (including, on occasion, the narration of myths), and undertook a search for a means to approach absolute truth—to know beauty, truth, and goodness in their pure, universal, and unmediated essence.

In order to support this quest, the difficulty of which was all too apparent even to Socrates, he placed it in a context that to our ears sounds more Indian than Greek, but that nevertheless represents his position. Socrates believed in metempsychosis—the survival of the soul after death via the process of transmigration and implantation of a soul in a new body. Thus Socrates argues, in *Meno* and elsewhere, that the process of gaining knowledge has a substantial component of recollection of knowledge held in previous existences. Indeed, the architecture of his approach to knowledge is quite similar to that presupposed by Buddhism, in which the task of penetrating to the essence of reality is so difficult that a number of lifetimes may be required. Thus that a human being realizes at all, that there is the possibility of grasping the pure forms of knowledge (or of enlightenment, within Buddhism) should provide impetus for the most strenuous efforts to achieve that perfect knowledge of being in a single lifetime, and to be released from the endless cycle of becoming.

Both Buddhism (and Yoga, for that matter) and Socratic phi-

Plato's Myths

losophy are concerned to overcome the dualism of existence
and our alienation from the single, true reality that is beyond
words. It is characteristic of enlightened knowledge in Buddhism that it cannot be put into words. Socrates was, however,
not so yielding on this point, and tested the limit of words to
express knowledge, reserving for "knowledge" the realm of
the truly real, all else being remanded to the category of "opinion." Plato's dialogues tell the story of this assault on ultimate
knowledge. *Meno* approaches the question of our geometric
intuitions by reference to recollection of knowledge held in a
previous existence. *Cratylus* is a critique of words and wordy
relations aimed at showing the tiresome fatuity of investigations which make clever guesses about what words mean by
inventing plausible etymologies for them. *Theatetus* urges
careful contemplation rather than being in a hurry—and points
out the self-destructive barbarism of specialized professional
knowledge-getting. *Parmenides* takes up the possibility of
knowledge (in Socrates' extreme sense) via natural scientific
investigations, and whether such knowledge can ever be more
than phenomenal appearance. The *Republic* takes up many of
the above issues in concert and develops the allegory of the
cave: that our opinions are like the shadows of real things cast
on the walls of a cave, and we are chained to our benches and
cannot turn to see the real world outside which, if we could
turn around, would at first blind us, and then puzzle us, for it
would not look like what we had known in the cave.

It is of course paradoxical that these dialectical investigations should, given Socrates' attitude toward writing, have
been written down. Indeed, it is to undo this paradox that
Plato puts so much of his own written thought in the form of a
dialogue spoken between Socrates and various contemporary
Athenians. Socrates deeply distrusted writing as a form of inquiry and communication, and was intensely discomfited with
the clear evidence that dialectic, especially when written
down, could be and in his lifetime was being used not for the
discovery of the true, but for material gain, for deception, and
for the manipulation of others in law and life. This distrust and

Natural Knowledge in Preclassical Antiquity

discomfort, which Plato shares, are the subject of the dialogue *Phaedrus,* in which Plato attempts to demonstrate why trying to do philosophy and discover truth by writing about it must fail. To accomplish this, and to show the spiritual danger of manipulative rhetoric (which does not attempt to speak truth, but only to win advantage), he introduces a physiological psychology based on direct wordless discourse through vision; a psychology which communicates beauty through the eyes, sets up qualitative changes in the body, causes the soul to grow and sprout wings, and creates conditions in which pairs of philosophic love-communicators may approach more closely to the knowledge of the real. In so doing Socrates (Plato) strives to provide a physiologic aim and basis for the proper pursuit of dialectical inquiry.

Because of the central importance of vision in this theory, I might have entitled this essay, "Plato's Optics." To have done so would not have been a mere conceit. Plato's cosmological dialogue *Timaeus* contains what has come to be known as "the Platonic Theory of Vision."[2] Plato's theory, which is material to our discussion here, is an extromission or projectile theory, in which seeing is not a passive reception of emanations from without, but an active projecting of seeing—in the form of visual rays—which somehow links up the object with the percipient. This is best represented by the cartoon image of the greedy child "eyeing" the pie cooling on the windowsill, with the "eyeing" graphically represented by a dotted line extending from the eye of the child to the pie. What is at issue here is an aspect of something that Alfred North Whitehead termed prehension—the act of grasping or fixing on a particular thing in a complex visual field—which captures the intuition of seeing as a complex act that goes far beyond the physiological mechanisms of the eye as a lens and the laws of optics to the intent and participation of the observer. Seeing is something that you do, not something that happens to you.

Indeed, Greek optical theory can be categorized broadly as consisting of three traditions. One is the medical tradition of the anatomy and physiology of the eye, and aims at under-

Plato's Myths

standing eye disease. The second is the tradition of mathematical optics, explaining perception of space in geometrical terms. The third is the physical, psychological, and philosophical analysis of the act of seeing as an intentional act.[3] Plato's theory of extromission of visual rays is this third sort of a theory. It generates questions that overlap with the other two notions of optics, but is in itself concerned only with establishing seeing as something that issues forth from a "seer."

So-called extromission or projectile theories of vision existed before Plato among the Pythagoreans. Alcmeon of Croton argued that the eye was possessed of a visual "fire" based on the experience that "when one is struck [this fire] flashes out."[4] What is described here is not a "gleam of anger" but the physiological fact that when you are struck in the eye, there is the subjective perception of a flash of light. That the perception is subjective, and a result of physical stimulation of the optic nerve without the objective production of light on the retina, was determined experimentally only in the nineteenth century.[5] But the characterization of the extromission theory as Plato's theory comes from the tremendous influence of the *Timaeus* in spreading it.

> The pure fire within us is akin to this [daylight], and they [the gods] caused it to flow through the eyes, making the whole fabric of the eye-ball, and especially the central part [the pupil] smooth and close in texture, so as to let nothing pass that is of coarser stuff, but only fire of this description to filter through pure by itself. Accordingly, whenever there is daylight round about, the visual current issues forth, like to like, and coalesces with it and is formed into a single homogeneous body in a direct line with the eyes, in whatever quarter the stream issuing from within strikes upon any object it encounters outside. So the whole, because of its homogeneity, is similarly affected and passes on the motions of anything it comes in contact with or anything that comes in contact with it, throughout the whole body to the soul and thus causes the sensations we call seeing. (Timaeus 45 b–c)[6]

Natural Knowledge in Preclassical Antiquity

Cornford calls this apparatus a "sympathetic chain"[7] and Lindberg correctly points out that Plato avoids a lot of geometrical problems with this notion in characterizing the act of seeing not as a simple projected signal that might have to be reflected back, but "the formation of a body through the coalescence of visual rays and daylight which serves as a material intermediary between the visible object and the eye."[8] This is a consolidated as opposed to a discrete mechanism of vision in which, without imputing this terminology to Plato, we can characterize the act of seeing as the distribution of a condition in a continuum, rather than a model of mechanical contact action of light or visual corpuscles, as in the (much) later optical theories of Newton or Descartes. We should recall that the propagation of light was taken by all physicists up to the time of Newton to be instantaneous.

Plato's theory of vision was subsequently incorporated into geometrical optics in Euclid's *Optica,* in the first postulate: "Let it be assumed that the rectilinear rays proceeding from the eye diverge indefinitely."[9] Though the subsequent development of Euclid's optical theory is purely geometrical, the intuition is of visual extromission. Our point here is not the intersection of Plato's theory of vision with geometrical optics, but that Plato postulates the ability of objects that intersect the visual medium to make physically mediated, but verbally unmediated, contact with the soul of the "seer." This is the subject taken up and employed in the *Phaedrus* in defense of the necessity that the philosophic pursuit of knowledge be limited to live dialectic. And it is here that we can develop a sense of why the dialectician and rationalist Plato makes serious use of myth, which he at other times invidiously characterized.

One of the most widely available and most recent English translations of *Phaedrus* is by Walter Hamilton, who introduces the dialogue thus: "The *Phaedrus* is chiefly valued by lovers of classics for its idyllic setting and its magnificent myth. It touches, however, on so many of the major themes of Platonic philosophy that it is unusually difficult to grasp its struc-

Plato's Myths

ture as a whole. Even in later antiquity it was debated whether it was primarily concerned with love or with rhetoric, or even with more general concepts such as the soul or the good or the beautiful."[10]

The idyllic setting is the banks of the Illisus outside Athens. Phaedrus is a young Athenian who has enticed Socrates outside the walls of Athens for a walk and a talk about the speech of the sophist Lysias. This is already an astonishing situation, as Socrates is a true *zoon politikon,* and tends to stay in the city: "the people in the city have something to teach me, but the field and the trees won't teach me anything" (230 c–d). The only other reference in the dialogues to Socrates venturing outside the walls is the opening lines of *Charmides,* when Socrates has returned from the Athenian army, which had taken severe punishment at the battle of Potidaea. I mention it because it is another side of Greek life always worth keeping in mind—the search for the good and the beautiful and the true in no way suggests to Socrates that he should not take his place in the phalanx amidst the screams and the dust and the blood. For Socrates, philosophy is not detachment but, like war, a matter of life and death.

The "magnificent myth" of the dialogue is the allegory of the soul as a charioteer and his two horses (one bad and one good horse) which, with no disrespect to Plato or Freud, we might liken, respectively, to Ego, Id, and Superego. It would not be the first case in which Freud tore a few sheets from Plato's book. The charioteer is the personal identity, the knowing consciousness, the Ego. The unruly horse is the Id that lunges forward oblivious to the whip and bit when he desires a destination. The trained and modest horse is the Superego, "constrained by a sense of shame," which keeps the charioteer from being tossed from his stand by the wildness of the bad "left-hand" horse (253).[11]

The "myth" also contains the fullest exposition of Plato's doctrine of rebirth. In contrast to some commonly described Indian versions of transmigration, souls in Plato's cosmic regimen go ten thousand years between reincarnations, and the

effects of goodness and badness in a previous life are not reflected in the character of the next rebirth, but rewarded or
expiated in the intermundium, after which souls elect the form
of their next rebirth. This process is accelerated only for those
who have sought the truth through philosophy. Only their
souls maintain continuity of knowledge of the real from one
life to the next, and only those who have sought rebirth and
the life of a philosopher three times running have hope of
release from the bondage of mortal existence (248–49). My
placing of the word "myth" in scare quotations at the beginning of the paragraph reflects my unwillingness to accept that
this is at all a fable, and reflects rather my conviction that the
charioteer and his horses are an allegory which Plato uses to
exhibit his fundamental conviction that this cycle of rebirth is
real. Plato's belief in the transmigration of souls is a much
underrated and fully theological component of his impetus
toward philosophical truth, and toward the ultimate aim of
seeking such truth: escape from mortal rebirth, and an eternal
existence as a disembodied soul in what one might well call
Nirvana. I would say that Plato's resort to this "magnificent
myth" is his way of saying what "goes without saying" in some
of the other dialogues: that knowledge of the truth has an aim
outside itself.

The late classical confusion about the dialogue's principal
aim—love? rhetoric? more general concepts?—already reflects
a loss of the metaphysical frame that defines the integrity of
the dialogue: namely, an analysis of the rhetoric of love to
exhibit the path to the good and the true and the beautiful, and
escape from the cycle of rebirth.

The speech of Lysias, which Phaedrus wishes Socrates to
hear, is a tissue of casuistry aimed at inducing a young man to
accede to the physical (bad horse) demands of a suitor, precisely because the suitor does not love the young man. This
speech, which we are led to believe was a commission from a
real suitor to the rhetorician Lysias, proceeds on the argument
that because love is a form of madness, the lack of love on the
part of the suitor is a positive aspect of the suit. It insures that

Plato's Myths

there will be no emotional confusion. There will be lust and reason, secret assignations balanced with public probity, but no importunity, no embarrassment, no loss of reputation.

If the whole point of Socrates accompanying Phaedrus on this walk has been to hear this speech of Lysias, there is a disagreement over how this is to be accomplished. Phaedrus admits that he has not got the speech by heart, but wishes nevertheless to deliver a summary of its main points in a speech of his own. Socrates insists to the contrary that he must hear the actual speech of Lysias, and that Phaedrus must produce the concealed manuscript of the same, and read from it. Phaedrus complies, and having read the speech summarized above asks Socrates for his reaction. Socrates replies that it was breathtaking—but the "it" turns out not to be the speech itself but Phaedrus's reading of it. Socrates does not trust the speech, but is moved by the intensity of Phaedrus's reading to "join his ecstasy" (234d). Phaedrus thinks Socrates is teasing, but he is not—quite. Moreover, in his subsequent analysis, Socrates indicates that the speech can be considered as three things: its form ("youthful exhibitionism; an attempt to demonstrate how he could say the same thing in two different ways"), its content ("the matter I didn't suppose . . . even Lysias himself could think satisfactory"), and the delivery by Phaedrus ("inspiring") (235).

Having heard the speech, Socrates does something quite unusual for him: he claims the capacity to do something well. He thinks that he could give a speech extempore on the same topic, inferior in neither matter nor form to that of Lysias. The image of capacity floats only for a moment, however, before Socrates returns to his chronic modesty: "Now I am far too well aware of my own ignorance to suppose any of these ideas can be my own. The explanation must be that I have been filled from some external source like a jar from a spring, but I am such a fool that I have forgotten how or from whom" (235c–d). Socrates, having made this boast, is then forced by Phaedrus actually to agree to deliver such a speech. Socrates coyly accedes but, rather curiously, insists he will "speak with my face covered; that way I shall get through the speech most quickly, and I shan't be put out by catching your eye and feeling ashamed" (237a).

Socrates, speaking with his face concealed, then delivers the first part of a speech designed to persuade a young man that the friendship of a suitor who does not love him is better than that of one who does. Socrates develops the argument that erotic desire is a form of madness in the lover, and when it finally passes, as it must, there will be great harm to the young man's character, reputation, and fortune. He will, in effect be left "seduced and abandoned." But at the very point when Socrates might have moved from this argument to the conclusion that the young man should yield to the "non-lover," he ceases his discourse abruptly and will not continue. Rather, he tells Phaedrus he has had a divine sign that his own speech thus far and the speech of Lysias which inspired it are both blasphemies "purchasing honour with men at the price of offending the gods" (242d). He must now, immediately, expiate the blasphemy, at his peril. Socrates then recalls for Phaedrus an old remedy: "When he lost his sight by speaking ill of Helen [of Troy], Stesichorus, unlike Homer, was sagacious enough to understand the reason; he immediately composed a poem . . . and as soon as he had finished what is called his palinode or recantation, he recovered his sight. Now I propose to to be even cleverer than our forbears; I mean to deliver a palinode to love before I suffer any harm for the wrong I have done him, and I will deliver it with my head uncovered, not muffling myself up from bashfulness as I did before" (243b).

With this device, Plato has now almost finished setting the stage for a presentation of his physiological and psychological theory of vision and true rhetoric. We should note that Socrates arranged things so that he and Phaedrus were not looking into each other's eyes, either when Phaedrus delivered the speech of Lysias, or when Socrates gave his own speech. In the first instance, Phaedrus was reading from a manuscript, and while Socrates was able to look at him, it was not possible for them to make eye contact. When it came Socrates' turn, Socrates

Plato's Myths

made the excuse of embarrassment, and spoke with his face hidden; Phaedrus could look at him, but not into his eyes. We should also note that when Socrates is compelled by a divine signal to cease his discourse, and to recant it, he does so under the threat of the loss of vision. The passage contains a not so subtle dig at Homer, whose blasphemies against Helen, by Socrates' account, left him permanently blind. Stesichorus was struck blind for blaspheming Helen, but realized his error and was able to regain his sight. Finally, Socrates has been able to discover his error before losing his eyesight, and is now prepared to make things right pre-emptively. The reason for all this will be apparent presently.

Socrates' recantation has two parts. The first is a dialectical analysis of the errors of his own previous speech and the speech of Lysias, into which is woven an argument on the immortality of the soul. The second embroiders the myth of the chariot of the soul and its attendant theme of soul-growth through love. These two parts are bound together in their joint necessity to explain the peculiar psychological and physiological conditions under which successful rebirth can be achieved by lovers devoted to philosophy.

In the first part of the recantation, Socrates locates the fundamental error of the earlier speeches in their unreflective repudiation of love in deeming it a form of madness. The cause of the non-lover was advanced by Lysias, and less emphatically by Socrates, on the grounds that if love is a form of madness, and if madness is a malady, then it is harmful to inflict it on the beloved (231). Socrates now wishes to divide the question and tabulate forms of madness, looking for the principle that links them. It then appears that there are several forms of "good madness." There is madness as the beneficent frenzy of prophetesses and oracles, who in their sane moments have no apparent gifts. There is madness as the warning of an underlying curse or possession, which can then be cured by prayer and purification. There is the madness of poetic possession by the muses (244–45). It is notable that Socrates has claimed to manifest all of them during the sequence of speeches: a voice

Natural Knowledge in Preclassical Antiquity

told him he could equal Lysias; during the speech he felt as if
he were possessed by the muses; and just afterwards an inner
voice warned of the danger his speech posed to his vision, and
provided recollection of the means to remove the curse.

It is true, argues Socrates, that love is a form of madness,
but it is closely connected to these other forms of good madness, and is, of them all, the best and most useful. The madness
of love is the recollection, by the immortal soul, of divine
beauty, a recollection potentiated by seeing earthly beauty. Of
all the pure forms known by the soul in heaven—justice, self-
discipline, and the others—only beauty retains its lustre here
on earth. The others are all too pale shadows of their true forms
to awaken the soul, even of those, falling into human rebirth,
who "retain sufficient memory" of the forms to see a glimpse
of them in their earthly counterfeits (250).

Beauty, in inspiring the madness of love, is our one link to
the world of the true forms, and our sense of sight is the means
by which this link is forged. Vision is the keenest of our physical senses, but yet even it cannot bring us knowledge: "what
overpowering love knowledge would inspire if it could bring
as clear [as clear as beauty] an image of itself before our sight."
Moreover, even when men are presented with the vision of
beauty, it is possible that earthly beauty will awaken not the
vision of absolute beauty, but mere animal sensuality.

Should the beholder of beauty get a glimpse of celestial
beauty through an uncorrupted view of beauty here on earth,
a view that avoids the quickening of base physical desire, he
sets in train an alternative physiological transformation that
yet begins with the same well-defined symptoms—a shiver
of dread, then adoration, replaced by heat and sweating—but
which here presage the reawakening and regrowth of his own
soul, and its capacity to ascend to heaven.

Socrates' portrayal of the renewed growth of the soul and
the reawakening of its capacity for flight is startlingly literal.
The soul in his description suggests a bird passing through
molt (Plato had, after all, elsewhere defined man as a feather-
less biped). As domestic fowl drop their feathers in the fall and

Plato's Myths

cease to lay eggs, so with the lengthening of the day in the spring do they regrow their feathers and come back into productivity. Here has man, falling into rebirth, molted and dropped the feathers of his soul, and languished in an unproductive wintery darkness. Now, beholding beauty, "he receives through his eyes the emanation of beauty, by which the soul's plumage is fostered, and grows hot, and this heat is accompanied by a softening of the passages from which the feathers grow . . . When in this condition the soul gazes upon the beauty of the beloved, and is fostered and warmed by the emanation which floods in upon it—which is why we speak of a 'flood' of longing—it wins relief from its pain [the itching and irritation of molt] and is glad" (251).

This conclusion of molt and nascent refledging transpires in the soul of the lover, but is only half the process with which Plato is concerned. For when "the beloved finds himself being treated like a god and receiving all manner of service from a lover whose love is true love and no pretense . . . his own nature disposes him to feel kindly towards his admirer . . . and when the loved one has grown used to being near his friend . . . the current of the stream which Zeus when he was in love with Ganymede called the 'stream of longing' sets in full flood toward the lover. Part of it enters into him, but when his heart is full the rest brims over, and as a wind or echo rebounds from a smooth or solid surface and is carried back to its point of origin, so the stream of beauty returns once more to its source in the beauty of the beloved. It enters in at his eyes, the natural channel of communication with the soul, and reaching and arousing the soul it moistens the passages from which the feathers shoot and stimulates the growth of wings, and in its turn the soul of the beloved is filled with love" (255).

There is an essential asymmetry to this process; it is the beauty of the beloved that causes the lover's soul to grow; but it is also the beloved's own beauty, reflected back from the surface of the brimming heart of the lover, that stimulates the soul of the beloved as well. There is a dialectic here, but it begins with a single truth (the beauty of the beloved), which

Natural Knowledge in Preclassical Antiquity

is then exchanged and exchanged again, its clarity growing with each iteration, as the awakened souls enlarge their capacity to see true beauty, potentiated by earthly beauty.

One might well ask why, if this is an optical theory, Socrates resorts to the image of a deflected wind or the echo of a sound to portray the rebound of beauty from the brimming heart of the lover back to the soul of the beloved. Why not just an optical reflection, or the analogy of a mirror? The reason for this follows immediately: "So now the beloved is in love, but with what he cannot tell. He does not know and cannot explain what has happened to him; he is like a man who has caught an eye infection from another and cannot account for it; he does not realize he is seeing himself in his lover as in a glass . . . He is experiencing a counter-love which is the reflection of the love he inspires" (255). The choice of the image of echo is in the interest of the ambiguity of that experience of hearing—often as if another had answered our call. The beloved sees not himself but the emanation of his beauty, and is confused as to its source. The upshot is that he falls in love.

It is by this process that the pair become lovers, and are faced with a choice. If they can restrain the physical release of their longing "in opposing to this impulse the moderating influence of modesty and reason . . . if the higher elements in their minds prevail," they can pass their lives in the pursuit of wisdom, and, in achieving self-mastery, put their souls at peace so that at the end of their lives "their wings will carry them aloft; they will have won the first of the three bouts in the real Olympic games [i.e., achieved the first of the three requisite lives as a philosopher necessary to escape rebirth altogether]" (256).

If the lovers, on the other hand, should get into their cups someday and, with their [moral] defences down "snatch at what the world regards as the height of felicity and . . . consummate their desire," they pass into the realm of the dead without wings, though they still avoid serving time beneath the earth. Those who transgress further than this, and seek physical intimacy without love, intimacy mingled with worldly calculation

Plato's Myths

and deception, give rise in their souls to an ignoble quality, and after death the souls so damaged will wander 9,000 years "devoid of wisdom" (270).

It is in this fashion that an extended argument for male homosexual pair bonding as the gateway to heaven stops short of becoming a brief for pederasty. Socrates urges lifelong attachments on the pattern of marriage with the stipulation that the more the physical desire for earthly beauty is forestalled, the greater the potential for soul growth—an arrangement colloquially described as a "purely platonic relationship." That the issue of soul growth should arise largely in the context of male homosexuality is one of the social peculiarities of Athenian life and had a determinative effect on the course of Socrates' philosophizing.

Socrates has proposed here a kind of psycho-physics. The term *soul* that I have used throughout is, of course, the common English translation of *psyche,* and it contains the sense of both "mind" and "soul." If the *psyche* is taken by Socrates to be immortal, it is nevertheless directly modified by sensory experience in life, both for good or for ill. In this case, the existence of a culture of male homosexual love provides an avenue to soul growth precisely because the soul *(psyche)* is susceptible to dynamic stimulation by an aesthetic quality— "beauty"—which reflects absolute beauty with an intensity that may awaken the soul. Socrates is very particular, as we noted above, about the strength of this dynamic stimulus, and its special capacity in forcing the soul to complete its molt-cycle in the course of a human lifetime. It is exactly (the image of the taming of the bad horse) in the taming of the homoerotic impulse that the strength of this psychic stimulus is directed toward the fledging of the soul.

Since it is this stimulus in this context which awakens the soul, Socrates urges in *Phaedrus* that the appropriate path to soul growth is the pursuit, by pairs of such lovers, of philosophic truth. This pursuit is, for Socrates (and Plato), the true aim of rhetoric. The art of rhetoric, like the art of medicine, has its technical, didactic counterpart—as indeed, spiritual prac-

Natural Knowledge in Preclassical Antiquity

tice has a theological counterpart. But for Plato, live philosophic discourse (true rhetoric) is a form of spiritual practice
that aims at self-development; therefore it requires practical
activity and in this respect resembles both medical practice
and meditation in search of enlightenment.

It is because of the power of the direct gaze in the process
of soul growth that Socrates has such concern in the dialogue
about where he and Phaedrus should have been looking during the blasphemous speeches. It would have been possible
to injure one or both of them had they looked into each other's
eyes during the delivery of insincere speeches with the power
to blind. The physiologic and psychologic effect is real, independent of the intention of the speaker, and this is the source of
Socrates' great unease about manipulative rhetoric in general,
about rhetoric for hire in particular, and about written speech
above all. Such rhetoric is bad for one's spiritual health, and
to the extent that it forestalls the soul growth which the practice
of a true art of rhetoric might foster, it thwarts the highest aim
of human existence.

The true art of rhetoric is also like the art of medicine, says
Socrates, in that "in both cases a nature needs to be analyzed,
in one the nature of the human body and in the other the nature
of the soul. Without this any attempt to implant health and
strength in the body by the use of drugs or diet, or the kind of
conviction or excellence you desire in the soul by means of
speeches and rules of behavior, will be a matter of mere empirical knack, and not of science" (270). That is, it will not be
science as *episteme* (true knowledge), rather than *praktike*
(practical knowledge in the political and ethical realm), or
poietike (poetical technique). Once it is agreed that the function of speech is to influence the soul (271), then the scientific
teacher of the art of speaking must understand the kinds of
speech, the kinds of souls, and the effects of the former on
the latter. Students of true rhetoric may then learn to fit their
treatment to the diagnosis and the prognosis of the souls which
come within the range of their art.

But like scientific medical practice, scientific rhetorical

Plato's Myths

practice is threatened by imposture and quackery, whereby patients with real and treatable ailments succumb to the ministrations and nostrums of charlatans to the profit of the charlatans. If the patients receive temporary relief from symptoms and some short-term benefit, they generally do so at the long-term expense of their health and sanity. Among the rhetorical quacks of Socrates' generation, the most potent, and therefore the most dangerous, were those who sold the patent medicine of written speech.

Socrates explains this danger to Phaedrus with recourse to another myth—that of the Egyptian god Theuth (Thoth), and his invention of the arts of number, calculation, geometry, astronomy, draughts and dice—and writing.[12] Theuth brings these arts to Thamus (Thamon—Ammon), king of the gods, and describing their utility, Theuth reserved special praise for writing: "Here O king is a branch of learning that will make the people of Egypt wiser and improve their memories; my discovery provides a recipe for memory and wisdom" (274e). Theuth urges that on this ground, writing, like the other arts, should be passed on to Egyptians in general. When Theuth presents this case for the general utility of writing, Thamus (Ammon) rebukes him sharply:

> Oh man full of arts, to one it is given to create the things of art, and to another to judge the measure of harm and of profit they have for those that shall employ them. And so it is that you, by reason of your tender regard for the writing that is your offspring, have declared the very opposite of its true effect. If men learn this, it will implant forgetfulness in their souls; they will cease to exercise memory because they rely on that which is written, calling things to remembrance no longer from within themselves, but by means of external marks. What you have discovered is not a recipe for memory, but for reminder. And it is no true wisdom that you offer your disciples, but only a semblance, for by telling them of many things without teaching them you will make them seem to know much, while for the most part they know nothing, and

Natural Knowledge in Preclassical Antiquity

as men filled, not with wisdom, but with a conceit of wisdom
they will be a burden to their fellows. (274e–275b)

Socrates' attitude toward writing, expressed in this myth—
which Plato may or may not have invented—has two clear foci.
One is the way in which writing, and reference to writing as
a means of calling things to remembrance, undermines the
growth of the soul, by weakening the memory. Where we find
memory *(mnemosyne)* in the much earlier writings of Hesiod
treated as a muse—a divine aid externally invoked, Socrates
and Plato transform it into an activity of the soul. This hypothesis is congruent with the notion that living discourse has a
special physiologic aim and foundation.

The second focus of Socrates' attitude toward writing is con-
sequent on Socrates finding himself in the midst of an unantici-
pated shift among rhetoricians to the written form. He sees a
generation of younger men who might have been and should
have been rhetoricians, orators, and philosophers turning into
speech writers, law writers, and lyric poets. Instead of writing
on their own souls and then pouring their wisdom into the
souls of others well prepared for the draught, they pour their
thought into a document as a vessel—a vessel from which anyone at all might drink: "Once a thing is put into writing, the
composition, whatever it may be, drifts all over the place, getting into the hands not only of those who understand it, but
equally of those who have no business with it" (275d–e). We
may note that, in the myth of Theuth, Thamus objects not to
writing per se, but to its being passed on to Egyptians in general. The Egyptians, whose reputation for wisdom among the
Greeks we have already encountered, left literacy in the hands
of a class of priestly scribes, and never mistook it for more than
a reminding tool. Socrates would clearly prefer this state of
affairs. Literacy, we might say, in the hands of the spiritually
unprepared, is a loose cannon, and writing for hire is more
or less like the irresponsible sale of automatic weapons—it
separates the technology from the social nexus which both
created and controlled it.

Plato's Myths

Authors who put forth the idea that written text can embody reliable and permanent truths, says Socrates, misunderstand the foundation of true knowledge and the means whereby it is imparted, namely, the recollection of real truths already known and hidden in the soul. Writing may remind one who already truly knows, but cannot teach one who does not know. Worse still, it imparts the illusion of knowledge and further diverts promising candidates from the kind of search which might bring about its recollection, and insure their good rebirth. In asserting that a text might hold a permanent truth, writers deliver the soul over to the dark hegemony of a dumb text, which can never respond to its inquiry—and substitute a dream image for the waking vision of the truth, closing off the world of the forms, and withering the soul.

It is above all in the criticism of writing that we come face to face with Plato's neo-Pythagorean beliefs. To treat this subject in detail would take us away from our consideration of myths in Plato's writings, but it is worth some brief comment. Pythagoras of Samos (fl. ca. 532 B.C.) founded a religious brotherhood of initiates who studied mathematics, practiced ethical vegetarianism, and pursued a course of spiritual discipline aimed at producing a good rebirth. After the death of Pythagoras, the movement rapidly split into two groups—the *mathematikoi* and the *akusmata*, the mathematicians and the hearers. The mathematicians continued the study of geometry with which Pythagoras's name is most closely associated, and continued to develop and teach a metaphysics of a universe constructed of number—of which our fullest account is Plato's *Timaeus*. The hearers, *akusmata*, were the religious adherents of the sayings of Pythagoras pertaining to ethical life, community existence, and rebirth. While the *mathematikoi* acknowledged the *akusmata* as true Pythagoreans, the reverse was not the case. From the standpoint of the *akusmata*, the *mathematikoi* were apostate in pursuing the study of mathematics and mathematical cosmology divorced from Pythagoras's particular spiritual quest.

The split between the writers and the students of true rhetoric in *Phaedrus* mirrors this Pythagorean schism. From the standpoint of Socrates, philosophical rhetoric absolutely requires the physical act of hearing under appropriate controlled circumstances within a group of initiates, not the promiscuous dissemination of techniques and texts. By definition, all *philosopoi* were *akusmata*—because the fundamental aim of philosophic rhetoric was the psycho-physical transformation, by hearing, of the souls of initiates already bound together by love, and therefore prepared for "refledging."

It is necessary to recall the neo-Pythagorean context of Plato's dialogues if we are to measure the natural-scientific content of these writings, and to understand Plato's attitude toward myth, at least in the *Phaedrus*. For Plato, *mythos* is a general term for all material not susceptible to dialectical treatment and demonstrative argument by the means of collection and division. Yet the *mythoi* in *Phaedrus* have distinct characters. The myth of the chariot is an allegory—an extended metaphor. The myth of the soul as a fledgling creature is an analogy. That is, it is a resemblance not in the objects of discourse (souls, fledgling creatures), but in their relations—their modes of growth. In modern developmental morphology, "analogous" organs in different classes of animals are those different in origin but similar in function. The theory of the physiological reception of beauty through the eyes is a description of an actual physical process, described by analogy with avian molt; what is described is the potentiation of the process which allows the organism (bird, soul) to develop the means to ascend into the heavens. In modern scientific terms, such a description would be called the avian model of the soul.

Finally, there is the *mythos* as a traditional tale—as in the case of the colloquy of Theuth and Thamus on writing. When Socrates told Phaedrus this myth, Phaedrus responded: "It is easy for you, Socrates, to make up tales from Egypt or anywhere else you fancy" (275b). Here we have Plato employing Phaedrus as an agent for the invidious characterization of myth as aetiological or moral fable. Socrates, normally a champion of dialectic, here retorts sarcastically that the Greeks of earlier days, "lacking the wisdom of you young people, were content in their simplicity to listen to trees or rocks, provided these told the truth" (275b–c). The message here seems to be that the truth of an utterance is not determined by its source. One should not necessarily disbelieve the truth of an assertion (that writing will harm the faculty of memory) just because it is presented as a fabulous utterance of a god, especially an Egyptian god, in whom one might disbelieve without hint of impiety. Neither should one believe an utterance, simply because it issues from the mouth of a skilled rhetorician. Over a wide range of the most important experiences, truth is discernible only by the awakened soul, and not by the technical manipulation of dialectical critique.

Certainly more important in the structure of the dialogue than this sort of tale is the *mythos* of the growth of the soul by the physiological reception of beauty through the eyes. Here Socrates has striven to provide a defense of live rhetoric against writing by developing an irreducible physiological component to true philosophical discourse. The awakening of the soul by contact with beauty, via what modern psychology might style a sublimated erotic impulse, is the only remaining window of opportunity to the perception and reception of the true forms by human perceptual apparatus. Philosophy is indeed like medicine, and a teacher like a physician of his pupils' souls. Socrates' anxiety over the loss of this avenue to soul growth, the most promising available to humankind, is well placed.

Suffice it to say that Socrates' focus on transmigration, his serious physiological psychology of the reception of the forms, and his scientific approach to vision are rarely considered in treatments of this dialogue, of which there are many.[13] I suspect the reason for this is that in the eyes of classicists devoted to the exegesis of Plato's dialogues, this material appears, scientifically speaking, "wrong." That is, they don't believe in transmigration, they don't accept the theory of physiological potentiation of soul growth by an infilling of substantive beauty emanations, and they reject the projectile theory of vision.

Natural Knowledge in Preclassical Antiquity

This has very clearly created a pressure to give all discourse on these matters in Plato's dialogues a figurative status as literary invention however "magnificent." Treating myth in Plato in an undifferentiated sense as fictional narrative—as allegory, fable, and trope—is a strategy, perhaps implicit, for enhancing the timeless character of the dialogues. The essence of the classical is after all the imputation of timeless truth. The implication is that this timeless veracity is marred or compromised or diminished by the presence of material which is, by positivistic standards, not true.

Yet this strategy, deliberate or implicit, falsifies the approach taken to rhetoric (and to mythic discourse) by Plato, and clouds the actual distinctions being made between mythic material and other forms of knowing. The distinction Plato makes here is not between false/figurative myth and true/representative rhetoric. Rather, the distinction is that between classes of things in the world not knowable by dialectic, and things which are knowable by dialectic. Rhetoric, especially in written form, is reduced, as it were, to a potent technology, and the principal problem Socrates addressed in this dialogue was how to induce his fellow Athenians to harness this powerful tool, and to employ it in a way which did not cause irreparable spiritual harm.

In fact, it may be that the principal injunction of the Hippocratic oath sworn by physicians, "to do no harm," is something Socrates would have urged on rhetoricians as well. Certainly his vision of the art implied a cohesive, self-policing guild of dedicated practitioners, with extensive memories, subtle diagnostic discrimination, and careful clinical training, and not a congeries of ambitious entrepreneurs, each armed with textbook and hungry for fees. As in medical practice, the science was to serve the art. In philosophy, and in medicine, when the art is subordinated to the science, the operation may be a success, and the patient may still die. This is the unifying theme of the dialogue of *Phaedrus,* and underlies all of Plato's approaches to the practice of philosophy.

▼ ▼ ▼

Notes

Chapter 1. Prehistory

1. Glyn Daniel, *The Idea of Prehistory* (Harmondsworth: Penguin Books, 1964), p. 10.

2. I use "men," "man," and "mankind" deliberately to suggest the outdatedness and onesidedness of this historical and mythological scheme. To use a phrase like "humankind" would modernize and to some extent validate this scheme, which I reject.

3. John Pfeiffer, *The Creative Explosion: An Inquiry into the Origins of Art and Religion* (New York: Harper & Row, 1982). For similar treatments of recent vintage, see John Gowlett, *Ascent to Civilization: The Archaeology of Early Man* (New York: Alfred A. Knopf, 1984); Randall White, *Dark Caves, Bright Visions: Life in Ice Age Europe* (New York: W. W. Norton, 1986); Grahame Clark, *World Prehistory in New Perspective*, 3d ed. (Cambridge: Cambridge University Press, 1977).

4. Glyn Daniel, *One Hundred Years of Archaeology* (London: Duckworth, 1950), p. 132.

5. See, e.g., André Leroi-Gourhan, *Préhistoire de l'art occidental* (Paris: L. Mazenod, 1965); Peter Ucko and Andre Rosenfeld, *Paleolithic Cave Art* (London: Weidenfeld & Nicolson, 1967); Ann Sieveking, *The Cave Artists* (London: Thames & Hudson, 1979); Margaret W. Conkey, *Art and Design in the Old Stone Age* (San Francisco: W. H. Freeman, 1982); and Henri Breuil's classic, *Four Hundred Centuries of Cave Art* (Montignac: Centre d'études et de documentation préhistoriques, 1952).

6. An extended discussion of this development may be found in Adam Kuper, *The Invention of Primitive Society* (London: Routledge, 1988).

7. See, especially, Leroi-Gourhan, *Préhistoire de l'art occidental.*

8. Niles Eldredge and Ian Tattersall, *The Myths of Human Evolution* (New York: Columbia University Press, 1982), p. 159.

9. Ibid., pp. 156–59.

10. Pfeiffer, *Creative Explosion,* p. 226.

11. The French, who have otherwise been so definite about much of prehistory, have employed the more agnostic designation *coup de poing,* which refers to any hand-held implement—including a derringer pistol, a set of brass knuckles (the *coup de poing américain*), or even an oil can.

12. Eileen O'Brien, "What Was the Acheulian Hand Ax?," *Natural History* 93 (1984): 20–23. When I contacted Dr. O"Brien in 1991 she was extending this line of research still further, though the results had not yet been published.

13. See George J. Kukla, "Loess Stratigraphy of Central Europe," in Karl Butzer and G. L. Isaac, eds., *After the Australopithecines* (The Hague: Mouton, 1975), pp. 99–188; George Kukla, "Pleistocene Land-Sea Correlations," *Earth Science Reviews* 13 (1977): 307–74; J. Cook et al., "Chronology of the Middle Pleistocene Hominid Record," *Yearbook of Physical Anthropology* 25 (1982): 19–65; Tag Nilsson, *The Pleistocene: Geology and Life in the Quaternary Ice Age* (Boston: D. Reidel, 1982).

14. S. A. Semenov, *Prehistoric Technology: An Experimental Study of the Oldest Tools and Artefacts from Traces of Manufacture and Wear,* trans. M. W. Thompson (New York: Barnes & Noble, 1964).

15. Lawrence H. Keeley, *Experimental Determination of Stone Tool Uses: A Microwear Analysis* (Chicago: University of Chicago Press, 1980).

16. Ibid., p. 76. See also Lawrence H. Keeley and Mark H. Newcomer, "Microwear Analysis of Experimental Flint Tools: A Test Case," *Journal of Archaeological Science* 4 (1977): 29–62.

17. Oswald Spengler, *Decline of the West* (New York: Alfred A. Knopf, 1926), p. 20.

Chapter 2. Egyptian Fractions

Epigraph: Lewis Mumford, *Technics and Civilization* (New York: Harcourt Brace & World, 1934), p. 275.

Notes to Pages 24–35

1. These examples courtesy of Prof. Thomas G. Palaima, of the University of Texas.

2. Quoted in Arthur Koestler, *The Sleepwalkers* (New York: Macmillan, 1959), p. 393.

3. Charles Piazzi Smyth, *Our Inheritance in the Great Pyramid* (London: A. Straham & Co., 1864).

4. Martin Gardner, "The Great Pyramid," in *Fads and Fallacies in the Name of Science* (New York: Dover, 1957), p. 176.

5. William Flinders Petrie, *The Pyramids and Temples of Gizeh* (London: Field & Tuer, 1883).

6. Ivars Peterson, "Ancient Technology: Pouring a Pyramid," *Science News* 125 (May 26, 1984): 327, reports Davidovits results from the 1984 Archaeometry Symposium at the Smithsonian.

7. Richard J. Gillings, *Mathematics in the Time of the Pharaohs* (Cambridge: MIT Press, 1972).

8. Ibid., p. 3.

9. For an extended discussion of these fractions and for information on traditional (pre-electronic) techniques of calculation, see A. A. Klaf, *Arithmetic Refresher for Practical People* (New York: Dover, 1964), esp. chap. 6 and pp. 99–102.

10. Gillings, *Mathematics*, p. 218.

11. Ibid., p. 227.

12. Karl Menninger, *Number Words and Number Symbols: A Cultural History* (Cambridge: MIT Press, 1969); German original, *Zahlwort und Ziffer* (1958).

13. Franz Boas, "Introduction to the International Journal of American Linguistics," *International Journal of American Linguistics* 1 (1917): 1ff.

14. See, e.g., Morris Kline, *Mathematics: The Loss of Certainty* (Oxford: Oxford University Press, 1980), chaps. 8–11.

15. Daniel C. Quinn, ed., *A Guide to Modern Mathematics* (Chicago: Science Research Associates, 1964), p. 40.

16. Otto Neugebauer, *The Exact Sciences in Antiquity*, 2d ed. (New York: Dover, 1957), p. 145.

17. Gillings, *Mathematics*, p. 20.

18. E. A. Wallis Budge, *An Egyptian Hieroglyphic Dictionary* (New York: Ungar, 1920), 1:414.

19. C. S. Peirce did not agree; see *Historical Perspectives on Peirce's Logic of Science: A History of Science*, ed. by Carolyn Eisele (New York: Mouton, 1985), "Egypt and Science," p. 334.

20. Gillings, *Mathematics*, p. 20.

21. Ibid., p. 105.

22. Hippolyte Ducros, "Étude sur les balances Égyptiennes," *Annales du Service des Antiquités de l'Égypte* 9 (1908): 32–53. For a discussion, see Bruno Kisch, *Ancient Scales and Weights: A Historical Outline* (New Haven: Yale University Press, 1965).

23. Sir William Flinders Petrie, *Ancient Weights and Measures* (London: Department of Egyptology, University of London, 1926), p. 7.

24. Ibid., p. 42.

25. B. L. van der Waerden, *Science Awakening: Egyptian, Babylonian and Greek Mathematics* (New York: Science Editions, 1963), p. 21.

26. Ibid., pp. 27–28.

27. Gillings, *Mathematics*, p. 154.

28. Ibid., p. 159.

Chapter 3. Hesiod's Volcanoes I. Titans and Typhoeus

Epigraph: G. S. Kirk, *Myth: Its Meaning and Functions in Ancient and Other Cultures* (Berkeley and Los Angeles: University of California Press, 1970), p. 10.

1. The standard critical edition with English notes is M. L. West, *Theogony*, edited with prolegomena and commentary (Oxford: Clarendon Press, 1966). Popular translations are N. O. Brown, *Theogony* (New York: Bobbs-Merrill, 1953), and Apostolos N. Athanassakis, *Hesiod: Theogony, Works and Days, Shield* (Baltimore: Johns Hopkins University Press, 1983); the bilingual [Greek/English] edition is H. G. Evelyn-White, *Hesiod, Homeric Hymns and Homerica* (Cambridge: Harvard University Press, 1936 [rev. of 1914, 1926], Loeb Classical Library no. 57). All modern editions are subsequent to the work of Alois Rzach, *Hesiodi Carmina* (Leipzig: B. G. Teubneri, 1902).

2. On the emergence of cosmic order, see G. S. Kirk, J. E. Raven, and M. Schofield, *The Presocratic Philosophers*, 2d ed. (New York: Cambridge University Press, 1983 [1st ed. 1957]), pp. 34–41; F. M. Cornford, *Principium Sapientiae: The Origins of Greek Philosophical Thought* (Cambridge: Cambridge University Press, 1952), pp. 194–95. For the relationship of invad-

ing Hellenic or Dorian religion and indigenous (chthonic) Min-
oan-Mycenaean religion, see Martin P. Nilsson, *A History of
Greek Religion*, 2d ed. (New York: Norton, 1964 [Swedish 1949,
English 1952]); see also Robert Graves, *The Greek Myths*, vol.
1, rev. ed. (New York: Penguin Books, 1960). For the notion that
cosmogonies are myths of sovereignty, see, e.g., J. P. Vernant,
The Origins of Greek Thought (Ithaca: Cornell University
Press, 1982 [translation of 1962 French original]), chap. 7.

3. On the date of the earliest texts of the *Theogony*, see West,
pp. 40–48.

4. Ibid., p. 382. See notes to lines 826, 827, 832, 839, 841,
846, 853, 860.

5. See Claude Lévi-Strauss, *The Savage Mind* (Chicago:
University of Chicago Press, 1966 [French original, *La pensée
sauvage*, 1962]). For the immediate and impoverishing literali-
zation of the linguistic metaphor behind much of his work, see,
e.g., Richard Macksey and Eugenio Donato, eds., *The Structuralist Controversy* (Baltimore: Johns Hopkins University Press,
1972).

6. The positivistic and formalistic nature of this analysis and
its tendency to arbitrarily exclude consideration of the content
of a myth are extensively discussed in Simon Clarke, *The Foundations of Structuralism* (Sussex: Harvester Press, 1981), pp.
157–209.

7. For the many developments and embroideries of this and
the battle with Typhoeus, see Joseph Fontenrose, *Python: A
Study of Delphic Myth and Its Origins* (Berkeley: University
of California Press, 1959), chap. 4, "Zeus and Typhon."

8. For an excellent summary of this interpretive material and
a clear account of research on the eruption, see Dorothy B. Vitaliano, *Legends of the Earth* (Bloomington: Indiana University
Press, 1973), pp. 179–271. Another fine scientific and historical
account is Fred M. Bullard, *Volcanoes of the Earth* (Austin:
University of Texas Press, 1973 [rev. ed. of 1962 orig.]), pp.
96–106. Scientific and archaeological evidence is compiled in
*Acta of the First International Scientific Congress on the Vol-
cano of Thera* (Athens: T.A.P. Service, 1971) and Christos Doumas, ed., *Thera and the Aegean World: Papers Presented at the
Second International Scientific Congress*, Santorini, Greece,
1978 (London: Thera & the Ancient World, 1978–80).

9. For someone who might admit it, see Eric A. Havelock, *The Literate Revolution in Greece and Its Cultural Consequences* (Princeton: Princeton University Press, 1982). See esp. the essay "The Preliteracy of the Greeks," pp. 185–207.

10. Tom Simkin et al., *Volcanoes of the World: A Regional Directory, Gazeteer, and Chronology of Volcanism During the Last 10,000 Years, Prepared by the Smithsonian Institution* (Stroudsburg, Penn.: Hutchinson Ross Publishing, 1981), pp. vii, 1–3.

11. Ibid., p. 21.

12. Ibid., pp. 37–41.

13. Ibid.

14. Ibid., p. 21.

15. See Dorothy B. Vitaliano and Charles J. Vitaliano, "Plinian Eruptions, Earthquakes, and Santorin, A Review," *Acta of the First International Scientific Congress on the Volcano of Thera,* pp. 88–98. "The Krakatoa eruption is the one most closely analogous to the Minoan eruption to the extent that it was an andesitic pumice eruption and resulted in the formation of a cal[d}era on the sea-floor" (p. 90n.).

16. This and the following description are from Bullard, pp. 85–91. His account digests those of B. G. Escher, *De Krakatau groep als vulkan* (Weltevreden, 1919), O. E. Stehn, *The Geology and Volcanism of the Krakatoa Group,* 4th Pacific Science Congress Guidebook (Batavia, 1929), and Howell Williams, "Calderas and Their Origin," *Bulletin of the Dept. of Geological Sciences* 25 (1941): 239–346.

17. Bullard, p. 88.

18. Vitaliano and Vitaliano, p. 95.

19. Ibid., p. 96. See n. 15.

20. C. Blot, "Volcanism and Seismicity in Mediterranean Island Arcs," pp. 33–44 in Doumas, ed., *Thera and the Aegean World.*

21. G. Komlos et al. "A Brief Note on Tectonic Earthquakes Related to the Activity of Santorini from Antiquity to the Present," pp. 97–107 in Doumas, ed., *Thera and the Aegean World,* pp. 103–4.

22. Bullard, p. 99.

23. See Bullard, pp. 99–100. See also Vitaliano and Vitaliano, p. 96; D. Ninkovich and Bruce Heezen, "Santorini Tephra"

in *Submarine Geology and Geophysics*, Colston Papers no. 17 (London: Butterworths, 1965), pp. 413–52; O. Mellis, "Volcanic Ash Horizons in Deep-Sea Sediments from the Eastern Mediterranean," *Deep Sea Research* 2 (1954): 89–92; L. Wilson, "Energetics of the Minoan Eruption," in Doumas, ed., *Thera and the Aegean World*, pp. 221–28; Jorg Keller, "The Major Volcanic Events in Recent Eastern Mediterranean Volcanism and Their Bearing on the Problem of Santorini Ash Layers," *Acta of the First International Scientific Congress on the Volcano of Thera*, pp. 153–61.

24. Robert Anderson et al., "Electricity in Volcanic Clouds," *Science* 148 (May 28, 1965): 1179–89.

25. West, p. 381.

26. Bullard, p. 337.

27. Ibid., pp. 234, 247, 343.

28. The most comprehensive treatment of the volcano in all its aspects is D. K. Chester et al., *Mount Etna: The Anatomy of a Volcano* (Stanford: Stanford University Press, 1985).

29. On the colonization of the region and the relocation from Naxos to Leontini, see *Cambridge Ancient History*, 2d ed. (Cambridge: Cambridge University Press, 1982), 3:3, "The Colonial Expansion of Greece," by A. J. Graham, pp. 83–162, esp. pp. 103–4.

30. For record of the earlier eruption, see Chester et al., p. 97 (table 3.3).

Chapter 4. Hesiod's Volcanoes II. Natural History of Cyclopes

1. G. S. Kirk, *Myth: Its Meaning and Functions in Ancient and Other Cultures* (Berkeley and Los Angeles: University of California Press, 1970).

2. See Claude Lévi-Strauss, *Structural Anthropology* (New York: Basic Books, 1963), chap. 11, on Oedipus; but for an outstanding exception, see S. C. Humphreys, *Anthropology and the Greeks* (London: Routledge & Kegan Paul, 1978), which contains both a history of the problem and informative studies in this mold.

3. G. S. Kirk, *The Nature of Greek Myths* (New York: Penguin, 1974).

Notes to Pages 75–88

4. Georges Dumézil, *Les Dieux des Indo-Européens* (Paris: Presses Universitaires de France, 1952), and *L'Idéologie tripartie des Indo-Européens*, vol. 31 (Brussels: Latomus, 1958); Victor Bérard, *Les Phéniciens et l'Odyssée* (Paris: Armand Colin, 1902), and *Did Homer Live?*, trans. Brian Rhys (New York: Dutton, 1931); F. Max Müller, *Lectures on the Science of Language* (New York: Scribner, Armstrong, 1869). See, for another source of nature-mythological interpretation, Adalbert Kuhn, *Die Herbakunft des Feuers und des Gottertranks* (Berlin: Dummler, 1859). The entire comparativist and unitist tradition is reviewed historically by C. Scott Littleton, *The New Comparative Mythology*, rev. ed. (Berkeley and Los Angeles: University of California Press, 1973). For theories in the Anglo-American tradition, see Marvin Harris, *The Rise of Anthropological Theory* (New York: Harper & Row, 1973), pp. 199–207.

5. Wilhelm Heinrich Roscher, ed., *Ausführliches Lexikon der greichischen und römischen Mythologie, 1884–1937*, reprint ed. in 10 vols. (Hildesheim: G. Olms, n.d.). For the flavor of Roscher's analyses, see his "Ephialtes: A Pathological-Mythological Treatise on the Nightmare in Classical Antiquity" and "An Essay on Pan," translated by James Hillman as *Pan and the Nightmare* (Zurich: Spring Publications, 1972).

6. Lucién Lévy-Bruhl, *La Mentalité primitive* (Paris: Librarie Felix Alcan, 1922).

7. Samson Eitrem, "Kyklopen," in Pauly-Wissowa *Real-Enzyklopädie der klassischen Altertumswissenschaft* (Stuttgart: J. B. Metzler, 1894–1919), 2327–47.

8. Ibid., 2343.

9. Ibid., 2343–45.

10. Kirk, *Myth*, p. 163.

11. Ibid.

12. George E. Mylonas, *Mycenae and the Mycenaean Age* (Princeton: Princeton University Press, 1966), pp. 34–37.

13. These can be seen as well in Alan J. B. Wace, *Mycenae: An Archaeological History and Guide* (New York: Biblo & Tannen, 1964), p. 136.

14. Robert Fitzgerald, *The Odyssey* (New York: Doubleday, 1961), p. 150.

15. Richmond Lattimore, *The Odyssey of Homer* (New York: Harper & Row, 1965), p. 142; Albert Cook, *The Odyssey* (New

York: Norton, 1974), p. 119; George Palmer, *The Odyssey of Homer* (Boston: Houghton Mifflin, 1892), p. 134. The correct translation is that of Gilbert Murray in *Odyssey* (Cambridge: Cambridge University Press, 1919): grain- or cereal-eating man.

Chapter 5. Thales and the Halys

1. W. K. C. Guthrie, *A History of Greek Philosophy*, vol. 1, *The Earlier Presocratics and the Pythagoreans* (Cambridge: Cambridge University Press, 1962), pp. 45–71; G. S. Kirk and J. E. Raven, *The Presocratic Philosophers* (Cambridge: Cambridge University Press, 1957), pp. 74–98; Hermann Diels, *Die Fragmente der Vorsokratiker* (Reinbeck bei Hamburg: Rowohlt, 1957), pp. 7–12.

2. Kirk and Raven, esp. pp. 84–86.

3. Guthrie, 1:39. Subsequent references are in the text.

4. Bruno Snell, *The Discovery of the Mind: The Greek Origins of European Thought* (New York: Harper & Row, 1953), chap. 9; Philip Wheelwright, ed., *The Presocratics* (New York: Odyssey Press, 1966), pp. 40ff; Benjamin Farrington, *Greek Science* (Harmondsworth: Penguin, 1944), chap. 2; Ernst Cassirer, *Language and Myth* (New York: Harper & Brothers, 1946); H. D. F. Kitto, *The Greeks* (Harmondsworth: Penguin, 1951); Jean-Pierre Vernant, *The Origins of Greek Thought* (Ithaca: Cornell University Press, 1982; French original, 1962). Historians of science seem more convinced of a cosmopolitan scientific culture in the eastern Mediterranean in the middle of the first millennium B.C. See George Sarton, *A History of Science*, vol. 1, *Ancient Science through the Golden Age of Greece* (Cambridge: Harvard University Press, 1953), pp. 160–98; B. L. van der Waerden, *Science Awakening: Egyptian, Babylonian, and Greek Mathematics* (New York: Science Editions [Wiley], 1963), pp. 82–90; Otto Neugebauer, *The Exact Sciences in Antiquity* (Providence, R.I.: Brown University Press, 1957), appendix 2, pp. 208ff. Even these writers insist that a distinct "logical turn" characterizes Greek thought over against its predecessors. For a recent pointed rejection of all "Greek Miracle" interpretations, see G. E. R. Lloyd, *The Revolutions of Wisdom: Studies in the Claims and Practice of Ancient Greek Science*,

Sather Classical Lectures, vol. 52 (Berkeley: University of California Press, 1987), p. 75 and passim.

5. Surendranath Dasgupta, *A History of Indian Philosophy* (Delhi: Motilal Banarsidass, 1975 [reprint of Cambridge University Press ed., 1922]), 1:28–59; Erich Frauwallner, *History of Indian Philosophy* (Delhi: Motilal Banarsidass, 1973 [reprint of the Salzburg ed. translation, Verlag Otto Muller, 1953–56]), 1:27ff; Mysore Hiriyanna, *Outlines of Indian Philosophy* (London: Allen & Unwin, 1932), pp. 48–86; Ninian Smart, *Doctrine and Argument in Indian Philosophy* (London: Allen & Unwin, 1964); Karl H. Potter, *Presuppositions of India's Philosophies* (Englewood Cliffs, N.J.: Prentice-Hall, 1963); I. M. Bochenski, *A History of Formal Logic* (Notre Dame, Ind.: Notre Dame University Press, 1961 [German orig., 1956]), pp. 461ff.

6. Guthrie, 1:29.

7. Quintus Septimus Tertullian, *De Carne Christi*, chap. 5. Cf. "Tertullianus," Pauly-Wissowa *Real-Encyclopädie der classischen Altertumswissenschaft* (Stuttgart: J. B. Metzler, 1894–1919),5a:822–44.

8. See, e.g., Alan B. Lloyd, *Herodotus: Book II, An Introduction* (Leiden: E. J. Brill, 1975), pp. 49–60, for a categorical denial of Egyptian influence on Thales.

9. Kirk and Raven, p. 78.

10. See Rhodes W. Fairbridge, ed., *The Encyclopedia of Geomorphology* (Stroudsburg, Penn.: Hutchinson & Ross), "Rivers: Meandering and Braiding," pp. 957–63; Luna B. Leopold and W. B. Langbein, "River Meanders," *Scientific American* (June 1966), reprinted in *The Physics of Everyday Phenomena* (San Francisco: W. H. Freeman, 1979), pp. 28–38; Luna B. Leopold and M. Gordon Wolman, "River Meanders," *Bulletin of the Geological Society of America* 71 (1960): 769–94.

11. See Karl A. Wittfogel, *Oriental Despotism* (New Haven: Yale University Press, 1955), esp. chap. 1, "The Natural Setting of Hydraulic Society," and chap. 2, "Hydraulic Economy." A more technical survey is R. J. Forbes, *Studies in Ancient Technology*, vol. 2 (Leiden: E. J. Brill, 1965), chap. 1, "Irrigation and Drainage." Similarly, see Norman Smith, *Man and Water: A History of Hydro-Technology* (New York: Charles Scribner's Sons, 1975). In the Chinese historical classic *Shu Ching* (5th century B.C.), the tribute to Emperor Yu highlights his legendary

status as the hero who tamed the waters and became the patron of hydraulic engineers. See Colin A. Ronan, *The Shorter Science and Civilization in China* (Cambridge: Cambridge University Press, 1981), p. 238. For a sense of the sophistication of this technology in classical times, see J. G. Landels, *Engineering in the Ancient World* (Berkeley & Los Angeles: University of California Press, 1978).

12. Richard D. Sullivan, "Historical Introduction," in William Coulson and Albert Leonard, *Ancient Naucratis*, in *Cities of the Delta*, American Research Center in Egypt Reports, vol. 4 (Malibu, Calif.: Undina Publications, 1981–82), 1:6–17. See also Lloyd, *Herodotus, Book II*, pp. 9–60.

13. Lloyd, pp. 14–23.

14. See Coulson and Leonard.

15. See H. R. Hall, "The Restoration of Egypt" and "Oriental Art of the Säite Period," in *Cambridge Ancient History*, vol. 3, *The Assyrian Empire* (Cambridge: Cambridge University Press, 1925), pp. 287–325.

16. T. F. R. G. Braun, "The Greeks in Egypt," *Cambridge Ancient History*, 2d ed., vol. 3, pt. 3, *The Expansion of the Greek World, Eighth to Sixth Centuries* B.C. (Cambridge: Cambridge University Press, 1982), pp. 32–57.

17. See Sidney Smith, "The Age of Ashurbanipal" and "Ashurbanipal and the Fall of Assyria," in *Cambridge Ancient History*, vol. 3, *The Assyrian Empire*, pp. 89–131, esp. chap. 4, sections 5–7.

18. Ovid, *Metamorphoses* VIII:162–66.

Chapter 6. The True Identity of Soma

Epigraph: Wendy Doniger O'Flaherty, "The Post-Vedic History of the Soma Plant," in R. Gordon Wasson, *Soma: Divine Mushroom of Immortality* (New York: Harcourt, Brace, Jovanovich, 1973), p. 95.

1. See Rig Veda 8.48, 9.113, 10.119. The best modern English version is that of W. D. O'Flaherty, *The Rig Veda* (New York: Penguin Books, 1981); a good standard scholarly edition is that of Karl Friedrich Geldner, *Der Rig-Veda aus dem Sanskrit ins Deutsche übersetzt und mit einem laufendem Kommentär Versehen* (Cambridge: Harvard University Press,

1951–57). For the Zend Avesta, see the translations of L. H. Mills, *The Zend Avesta*, pt. 3 (Oxford: Clarendon Press, 1887), vol. 31 of *The Sacred Books of the East*, ed. F. Max Müller; see esp. Yasna 9 and Yasna 10.

2. Robert C. Zaehner, *The Dawn and Twilight of Zoroastrianism* (London: Weidenfeld & Nicolson, 1961), p. 66.

3. Jeffrey H. Tigay, *The Evolution of the Gilgamesh Epic* (Philadelphia: University of Pennsylvania Press, 1982), pp. 214ff; Alexander Heidel, *The Gilgamesh Epic and Old Testament Parallels*, 2d ed. (Chicago: University of Chicago Press, 1949), pp. 141ff; B. S. Childs, "Tree of Knowledge," in *The Interpreter's Dictionary of the Bible* (Nashville: Abingdon Press, 1976), pp. 695ff.

4. Monier Monier-Williams, *Religious Life and Thought in India: Vedism, Brahmanism, Hinduism* (New Delhi: Oriental Books Reprint Corp., 1974; reprint of London: John Murray, 1883), pp. 12–13.

5. O'Flaherty, "Post-Vedic History," p. 146.

6. Ibid., p. 147.

7. Ibid., p. 104.

8. Ibid., p. 104; reference is to William Roxburgh, *Flora Indica; or, Descriptions of Indian Plants* (New Delhi: Today and Tomorrow's Printers and Publishers, 1971; reprint of Carey's ed., Seranpore, 1832), p. 33.

9. Rajendra lala Mitra, "Spiritous Drinks in Ancient India," *Journal of the Asiatic Society of Bengal*, part 1, no. 1 (1873), 1–23.

10. O'Flaherty, "Post-Vedic History," p. 111.

11. Robert C. Zaehner, *Mysticism Sacred and Profane* (Oxford: Clarendon Press, 1957); Aldous Huxley, *The Doors of Perception* (London: Chatto & Windus, 1954).

12. Varro E. Tyler, "The Physiological Properties and Chemical Constituents of Some Habit-Forming Plants," *Lloydia* [now *Journal of Natural Products*] 29 (1966): 275–92, p. 285.

13. Ibid.

14. Wasson, *Soma*, p. 3.

15. Zaehner, *Dawn and Twilight*, p. 90. See also Herman Lommel, *Die Yasts des Avesta* (Göttingen: Vandenhoeck und Ruprecht, 1927), and Fritz Wolff, *Avesta: Des Heiligen Bücher der Persen* (Strassburg: K. J. Trübner, 1910).

16. Zaehner, *Dawn and Twilight*, p. 85.

17. Julius Eggeling, trans., *The Satapatha Brahmana* (Oxford: Clarendon Press, 1881 [pt. 1], 1884 [pt. 2], 1894 [pt. 3]); published in *Sacred Books of the East*, ed. F. Max Müller, vol. 11 (pt. 1), vol. 26 (pt. 2), vol. 41 (pt. 3). See sect. 5.1.2.10.

18. Lester Grinspoon and James Bakalar, *Psychedelic Drugs Reconsidered* (New York: Basic Books, 1979), pp. 28–29.

19. Ara der Marderosian, "Hallucinogenic Indole Compounds from Higher Plants," *Lloydia* [now *Journal of Natural Products*] 30 (1967): 23–28, p. 23.

20. Braja Lal Mukherjee, "The Soma Plant," *Journal of the Royal Asiatic Society of Great Britain and Ireland*, 3d series (1921), p. 241, cited in O'Flaherty, "Post-Vedic History," p. 128; Jogesh Chandra Roy, "The Soma Plant," *Indian Historical Quarterly* 15 (June 1939): 197–207.

21. O'Flaherty, "Post-Vedic History," pp. 128–129.

22. *CRC Handbook of Chemistry and Physics*, 68th ed. (Boca Raton, Fla.: CRC, 1987–88), pp. c-354–55.; *Merck Index*, 8th ed. (Rahway, N.J.: Merck, 1968), pp. 765–66.

23. A. K. Nadkarni, *Indian Materia Medica* (Bombay: Popular Book Depot, 1954), 1:904.

24. Wasson, *Soma*, p. 37. Subsequent references are in the text.

25. Lysergic acid amides have been proposed by Carl A. P. Ruck, an associate of Gordon Wasson and Albert Hofmann, as the active ingredient in the potion ingested by participants in the Eluesinian mysteries. His argument is suggestive but does not go beyond plant morphology nor does it explore the techniques of preparation. His priority in making the (independently developed) suggestion of ergot as the source of hallucinogens in antiquity must be acknowledged, though he makes no mention of Soma. See R. Gordon Wasson, Carl A. P. Ruck, and Albert Hoffmann, *The Road to Eleusis* (New York: Harcourt Brace Jovanovich, 1978).

26. Clyde M. Christensen, *Molds, Mushrooms, and Mycotoxins* (Minneapolis: University of Minnesota Press, 1975), p. 35.

27. Ibid., pp. 36–37, 39.

28. Ibid., p. 37.

29. Ibid., p. 38.

30. Trevor Robinson, *Biochemistry of the Alkaloids* (New York: Springer Verlag, 1968), p. 77.

31. Egil Ramsted, "Chemistry of Alkaloid Formation in Ergot," *Lloydia* 31 (1968): 327–41, p. 327.

32. Albert Hofmann, *LSD: My Problem Child* (New York: McGraw-Hill, 1980), p. 12. Subsequent references in this section are in the text.

33. R. G. Wasson, "Notes on the Present Status of Ololiuqui and Other Hallucinogens in Mexico," Harvard University *Botanical Museum Leaflet* 19 (1963): 137–62.

34. Hofmann, *LSD*, p. 123.

35. Ramsted, p. 331; A. Stoll and A. Hofmann, "The Chemistry of the Ergot Alkaloids," in S. W. Pelletier, ed., *Chemistry of the Alkaloids* (New York: Van Nostrand Reinhold, 1970), p. 279.

36. F. R. Allchin, "Early Cultivated Plants in India and Pakistan," in Gregory Possehl, ed., *Ancient Cities of the Indus* (Durham, N.C.: Carolina Academic Press, 1979), pp. 249–52, p. 250. A longer version of this paper appears in Peter J. Ucko and G. W. Dimbleby, eds., *The Domestication and Exploitation of Plants and Animals* (London: Duckworth, 1969), pp. 323–29.

37. Allchin, p. 249.

38. Colin P. Masica, "Aryan and Non-Aryan Elements in North Indian Agriculture," in Mahdev M. Deshpande and Peter Edwin Hook, eds., *Aryan and Non-Aryan in India* (Ann Arbor: Center for South and Southeast Asian Studies, University of Michigan, 1979), pp. 55–151, p. 104.

39. Allchin, p. 251.

40. Masica, pp. 103–4.

41. Ibid., p. 105.

42. E. B. Havell, "What Is Soma?," *Journal of the Royal Asiatic Society of Great Britain and Ireland*, 3d series (1920), pp. 349–51, p. 351.

43. Robinson, *Biochemistry*, pp. 3–4.

44. Christensen, *Molds*, p. 42.

45. Ibid.

46. Robinson, p. 5.

47. Wasson, *Soma*, p. 36.

48. Richard Schultes and Albert Hofmann, *The Botany and Chemisty of the Hallucinogens* (Springfield, Ill.: Charles Thomas, 1973), p. 36. In Africa it grows with oak and eucalyptus; in the north temperate zone, in overgrown pastures.

49. Christensen, *Molds,* pp. 46–47.

50. *Merck Index* 1968, p. 414.

51. See Nadkarni, *Indian Materia Medica,* pp 1198–99.

52. Ibid., p. 723.

53. Satapatha Brahmana 5.3.3.4, quoted by Eggeling, Introduction, p. 1.

54. Havell, "What Is Soma?," pp. 349–51.

55. Nadkarni, p. 324.

Chapter 7. Plato's Myths

1. Plato, *Phaedrus.* Translations here are from the version of Walter Hamilton, *Plato: Phaedrus and Letters VII and VIII* (New York: Penguin, 1973). Numbers are to standard divisions of the original. Here, 229d-e; subsequent references in text. There is a more recent translation with commentary by C. J. Rowe, *Plato's Phaedrus* (Warminster: Aris & Phillips, 1986).

2. David C. Lindberg, *Theories of Vision from Al-Kindi to Kepler* (Chicago: University of Chicago Press, 1976), p. 3. See also Wilbur C. Knorr, "Archimedes and the Pseudo-Euclidean Catoptrics: Early Stages in the Ancient Geometric Theory of Mittors," *Archives Internationales d'histoire des Sciences* 35 (1985): 28–105.

3. Ibid., p. 1.

4. Theophrastus on Alcmeon, quoted in Lindberg, p. 4.

5. H. von Helmholtz, "Recent Progress in the Theory of Vision"[1867], in Richard M. amd Roslyn P. Warren, eds., *Helmholtz on Perception* (New York: Wiley, 1968), p. 85.

6. Plato, *Timaeus,* in Edith Hamilton and Huntington Cairns, eds., *The Collected Dialogues of Plato* (New York: Bollingen Foundation, 1961), trans. Benj. Jowett.

7. F. M. Cornford, *Plato's Cosmology* (New York: Humanities Press, 1952), p. 153, n. 1.

8. Lindberg, p. 5.

9. Quoted by Lindberg, p. 12, from Morris R. Cohen and I. E. Drabkin's *A Source Book in Greek Science* (Cambridge: Harvard University Press, 1958), p. 257.

10. Hamilton, *Phaedrus,* Introduction, p. 7.

11. The sexual character of much of the dialogue and its Freudian interpretation are taken up in G. R. F. Ferrari, *Listen-*

ing to the Cicadas: A Study of Plato's Phaedrus (Cambridge: Cambridge University Press, 1987); see esp. chap. 5, "Myth and Understanding."

12. The apparently anomalous presence of the names of games of amusement in a list of noble inventions can be explained by the necessity of mastering the calculation of chance—of odds—in order to play such games successfully.

13. See Ferrari, *Listening to the Cicadas*, for a full bibliography of studies of the dialogue.

Index

Index

Index

Index

Designed by Glen Burris
Set in Caledonia text and Caslon 540 display by
Maryland Composition Company, Inc.
Printed on 55-lb. Sebago Antique Cream and bound in ICG Arrestox cloth by
The Maple Press Company